NOT JUST FOR THE BOYS

ATHENE DONALD

NOT JUST FOR THE BOYS

Why We Need More Women in Science

Great Clarendon Street, Oxford, OX2 6DP,
United Kingdom

Oxford University Press is a department of the University of Oxford.
It furthers the University's objective of excellence in research, scholarship,
and education by publishing worldwide. Oxford is a registered trade mark of
Oxford University Press in the UK and in certain other countries

Published in the United States of America by Oxford University Press
198 Madison Avenue, New York, NY 10016, United States of America

British Library Cataloguing in Publication Data
Data available

Library of Congress Control Number: 2023930747

ISBN 978–0–19–289340–6

DOI: 10.1093/oso/9780192893406.001.0001

Printed by Integrated Books International, United States of America

To my granddaughters

PREFACE

This book is written for anyone who is wondering why, in spite of decades of effort to promote change, the numbers of women pursuing careers in the physical sciences and engineering still remain small and the numbers of women reaching the top of biomedical research are not at all in proportion to those who start out. Despite barriers appearing to have been removed, less visible hurdles remain to trip up many women.

Some of the answers to these questions are subtle, but many are not. Somehow society is still stuck in a time warp, where women are expected not to get their hands dirty on a construction site or labouring at a lab bench. This view appears to continue to hold, despite many recent examples of women making a difference, something particularly noticeable during the Covid-19 pandemic, where the role of women in developing vaccines and contributing to public health has been so prominent. Such views of what women can and should do are outdated and need to change, if society is to benefit from all they have to offer in the scientific domain.

We may have progressed beyond the 19th century belief that tackling complex mathematics would damage a woman's reproductive system, and the mid-20th century view leading to a Princeton graduate remarking, as the long-established institution considered admitting women, *Keep the Damned Women Out*.[1] Yet the repercussions of this history can still be felt across the so-called STEM (Science, Technology, Engineering, and Mathematics)[2] subjects, where women remain in a very noticeable minority.

This book explores the obstacles that women face in studying and practising these different scientific and technical disciplines today, and how things can be improved. It is not a book written solely for women, to help them understand the hurdles they face or might face if they enter the STEM professions. Crucially it is also written for men to read and consider their own actions: how these may be influencing the women they work with, what they might do to improve the work environment for all, and how they personally can support women's progression. Furthermore, it is not a book written solely for the practising or would-be scientist, but also for parents, policy-makers, and employers, whose decisions impact on how girls make disciplinary choices from an early age and what atmosphere they subsequently encounter at university and in the workforce.

I hope the next generation of would-be female scientists don't continue to face the same obstacles many do today. Around the world, we can do better. Society will be the stronger for it if we welcome these women and encourage them into the scientific world. Diversity improves outcomes, as business has begun to recognize. It is time for our laboratories, schools, and industries to do the same and ensure that their workforce reflects this reality.

ACKNOWLEDGEMENTS

This book is about a topic that I have had plenty of opportunity to think about, both experientially and through my roles on different committees and in different organizations. During the course of this work, I have interacted with many wonderful and helpful people who have taught me so much. To all of you, thank you. I can't name you each individually, but I will highlight some who have been particularly stimulating, challenging, and/or informative. So, thank you to Jocelyn Bell Burnell, Beth Bromley, Julia Buckingham, Ruth Cameron, Jane Clarke, Hannah Devlin, Sigrid Fisher, Mark Geoghegan, Val Gibson, Helen Gleeson, Dot Griffiths, Vivien Gruar, Esther Haines, Julia Higgins, Jenny Higham, Sheila MacNeil, Nancy Lane Perham, Hilary Lappin-Scott, Ottoline Leyser, Miriam Lynn, Margaret Mackie, Rachel Oliver, Rachel O'Reilly, David Peet, Gwen Reilly, Tony Ryan, Jeremy Sanders, Jim Smith, Gwyneth Stallard, Barbara Stocking, Sarah Teichmann, Paul Walton, Tom Welton, Isabelle Vernos, Lesley Yellowlees, and Andrew Webber. I would also like to thank Catherine Sutherland, who introduced me to Mary Astell and allowed me to examine some of her books and annotations and set me straight on what I wrote.

I am deeply grateful to all who not only have discussed these matters with me but looked at all or parts of the book, from their very different perspectives, including Melinda Duer, Patricia Fara,

Uta Frith, Mark Goldie, Richard Jones, Margaret Mackie, Gina Rippon, and Meg Staff. I am also deeply grateful to my editor Latha Menon at OUP and their anonymous readers who all helped the book to gain polish and focus. Finally, I owe an enormous debt to my husband Matthew, who has patiently listened to me—and sometimes not quite so patiently—talk about this topic repeatedly over the years, as well as constantly proffering emotional and practical support throughout our long-lasting partnership.

CONTENTS

CHAPTER 1

WHAT'S THE PROBLEM?

I still miss the buzz of hands-on research as an experimental physicist. Back in the days before everything became digital, I would spend hours sitting in the dark, staring at the screen of an electron microscope. Maybe this isn't everyone's idea of fun, but I spent a significant part of my early 20s doing just that, full of excitement about what new insights each session on the electron microscope might bring. There is a sense of wonder in staring at something which has never been seen by anyone previously. It isn't like the aesthetic contemplation of Renaissance Art: it's not about inherent beauty. The wonder arises from the unknown, the previously unknowable, which may suddenly open up. Or it may not. For every successful hour on the microscope there will be at least as many, sometimes ten times as many, when nothing goes quite right and there's essentially nothing to see. That's what research is like: an exhilarating, exhausting roller coaster of emotions.

The period of my research life that was most heady, and pumped the adrenaline round my body fastest, occurred when I was a young researcher in the USA. I was in upstate New York, at Cornell University, where I was working as a postdoctoral researcher ('postdoc'). I am at heart a microscopist; I have spent many hours staring—with joy, frustration, and not infrequent boredom—down microscopes. When my research is going well, every new sample I put in front of my eyes I look at with hope and excitement, forever wondering what new feature I am likely to see and how it will fit into the jigsaw that I am trying to construct of facts, theories, my prior observations, and other people's evidence. The adrenalin hit is overwhelming when things work. Unfortunately, many periods of research are anything but like that, but the memory of that buzz remains as a permanent feel-good moment.

This period that produced the most sustained research 'high' was my second postdoctoral appointment. After a very unproductive couple of years, I changed fields—from researching metals to plastics—and also my professor, but stuck with working with microscopes, mainly electron microscopes. I was lucky. I was able to overcome the hurdle that my first, extremely unsuccessful postdoc posed, and continue a research career in academia. Many researchers do not have that luck. Was there a gender angle in that first failure? Quite possibly, since I was the first woman to hold a postdoc in the entire department (which was part of the Engineering Faculty), but I can't be sure. Many women may be much more certain of gender playing a role in their discomfort at any stage in their careers, a statement as true in science as in any other. What is different is that the number of women who even start a course at university in STEM (I will frequently use 'science' loosely to cover all the different STEM disciplines for convenience) is typically well below 50%, the exception being in the biological sciences, so that losses higher up the pipeline have a stark effect.

Why do women not prosper in the scientific workforce? Why are girls discouraged from ever embarking on science at university in the first place, let alone beyond? This book will look back briefly at how, historically, society has always excluded women from the scientific sphere and discourse, and how far we have come. It will explore societal expectations during both childhood and working life. It will bring together quantitative evidence from some of the latest studies of the systemic disadvantages under which women operate in academia around the world with the developing science of how our brains are—and more importantly aren't—gendered. It will look at the societal challenges women in the workforce face and how they play out within the scientific sphere. Only if we understand the present can we realistically take steps to improve the future.

The book will discuss how scientific research is done in practice, in order to dispel some common myths. Myths such as that science is not creative or that it is done by a lone genius in an ivory tower. Modern science is not like that at all, and these myths can be very off-putting to many sections of the population. A better appreciation of the collaborative, creative, and multi-disciplinary nature of science is likely to lead to its appeal to a far wider swathe of people, maybe especially to women, who are often brought up to be team-players.

The reality is, however, that men and women typically experience the workplace differently as they progress along their scientific career paths, with women encountering problems that aren't always even visible to men. Let me give you one striking example of negativity experienced by a woman in science, someone right at the top of their game: the German Nobel Prize winner Christiane Nüsslein-Volhard. You might think, if you've just been awarded the Nobel Prize in one of the sciences, that the world will want to celebrate with you. Well, maybe not. Not, at least, if you're a woman. So said Nüsslein-Volhard in an interview a few years back with the newspaper, *der Spiegel:*

> *After a while you get used to being a Nobel Prize laureate. But some colleagues couldn't bear that I got the prize. I remember the day the news came out: I called the managing director of the institute and said: I got the Nobel Prize and we have to have a party.*[1]

The interviewer followed up with '*How did he react?*' to which she replied:

> *He said: 'Can you please organize the champagne yourself? I have no time to take care of that.'*

A woman is expected to do the 'housekeeping', something we will see more of later in the book.

As a female scientist you don't have to be a Nobel Prize winner to recognize that how men react to women's success is not necessarily identical to their response to men's achievements. I had a colleague who, on being elected to the Royal Society (the UK's national academy of science), was told her department would not be celebrating this because it 'wasn't her turn'. In my case, I had a professor write to me on my own election saying in a distinctly barbed tone '*you never know how clever colleagues are around you until they make FRS*'.

At junior levels, too, women scientists can experience the equivalent sort of negativity. As Nüsslein-Volhard puts it in the same interview:

> *My PhD supervisor was very charming and I certainly thought he could be a mentor and guide me. But he couldn't bear that a student was better than him. In the end, he tried to get in my way. I'm still bitter about it today.*

Although she was a student many years ago, I fear some women would still have supervisors who react the same way. It is not a million miles away from the challenge faced by female leaders, such as Hillary Clinton during her 2008 and 2012 election campaigns: likeability versus competence. This is not a tension that men, in any sphere, seem to be up against.

'The world needs science and science needs women', says the L'Oreal strapline. They are, in their own way, trying to remedy some of the problems by providing funding for gifted female scientists from all round the world, but they can only scratch at the surface of the problems that still beset women progressing. *'9% is not enough'*, headlines the British Institute of Engineering and Technology's campaign for gender equality, reflecting the extreme paucity of women entering the engineering profession. A mere 9%! Why does it matter? Why should *you* care that women fall by the wayside at every stage of a scientific career?

Firstly, because it might be your talented daughter, granddaughter, god-daughter, niece, neighbour's daughter ... it might be the next Marie Skłodowska Curie or Rosalind Franklin who gets put off or actively shunted out. Their brilliance may never be able to deliver if at 15 an unthinking teacher implies that engineering is not very lady-like and wouldn't they rather study English literature at university? The days of recommending a secretarial course for the smart girl may be long past, but similar stereotyping may still occur about career options, directly or indirectly. If a teacher implicitly believes girls are less good at maths, for instance, they will underperform and thereby rule themselves out of career tracks requiring maths skills.[2] Schoolchildren who are not provided with appropriate careers advice are not likely to know they might want to enter a specific sector. Construction, for instance, in the UK at least, is overwhelmingly male. If a 2017 survey by a major construction company[3] is representative, this trend is likely to continue: it found that, of children that had been offered careers advice, 40% of boys were told about construction, compared with only 29% of girls.

Secondly, our society globally is facing a range of existential crises, including climate change, the spread of diseases from Ebola to Covid-19, and food security, and their solutions all have science and scientists sitting firmly at their heart. We, the public, are often told we need more scientists to tackle these problems, to innovate, to stimulate the economy and to push up productivity, yet societies equally often convey messages that discourage girls from thinking about science as a career. We are potentially losing half our talent if we let this occur, to everyone's detriment. As the Mexican biologist Esther Orozco, the L'Oreal/UNESCO Laureate 2006 for Latin America and the Caribbean, put it: '*When a talented woman is led away from science, humanity loses half of its talent and much more of its sensitivity and intuition*'.[4]

Who is to provide the innovation and insight to overcome these massive obstacles to life as we know it, wherever we may live? We cannot afford to keep (or actively drive) women out of science if the global solutions to the problems we face are to be confronted and overcome. It's not simply that we need more scientists and technologists, although we do; it is that we need different ways of thinking about the problems. Just as in a business context, where the diversity of teams is known to lead to more innovative solutions yielding ultimately bigger profits, so diversity of approaches in science can yield novel insight and originality. Diversity covers all kinds of strands, but what it most certainly does not mean is a group looking just like their boss or professor.

Gender is just one form of diversity but, given that women are approximately half the population, it makes sense to worry if a substantial proportion of that half are lost to the push for innovation and originality due to the way our society works. They are, of course, not the only part of the population which has systematically been disadvantaged over the years. Diversity in our research approaches means encouraging all these segments of society and making sure they are welcomed into the scientific fold. Ethnic background, sexual orientation, and disability all may lead to significant disadvantages arising from cultural attitudes which we need to root out of our labs as across our society; intersectionality will make every problem worse. However, this book's primary focus will remain on the issue of women, the area I am most familiar with, while recognizing the importance of these related issues.

My aim in this book is to bring together recent evidence from a wide range of different spheres, including neuroscience and the social sciences, to construct a coherent and integrated picture of the current situation for women in science. Anecdotes and quotes from practitioners of science, both male and female, about how they approach their science will serve to illustrate some of the hurdles and how they can and should be tackled. In my previous role as the University of Cambridge's Gender Equality Champion, I learned an enormous amount about the specific issues facing women in my own university.[5] And having interviewed many senior scientists from different backgrounds, formally and often informally,[6] I have plentiful examples, to add to my own experiences, of how different women have faced up to the challenges and whether or not their approaches have worked.

Even the most successful scientist has setbacks, but these can feel particularly painful if they arise because of gender and not personal inadequacies, as anecdotes to be found in this book and elsewhere show. There are quotations from Nobel scientists in the pages that follow, but from others at widely different career stages too: very few people make it right to the top of the ladder. It is important not to use Nobel winners as 'typical' examples of what a scientist looks like and what they have experienced, because inevitably they will be anything but.

Attitudes matter in how we interact with our children, but also in how we judge people later in their lives. When it comes to gender bias, let me give you a striking example of a woman scientist's experience and the equivalent control experiment for a man. In case you are wondering how you can provide an appropriate male control to match against a woman, the answer is—find a transgender scientist. The late American neurobiologist Ben Barres, born Barbara, is a striking example. I find his story compelling as evidence that bias against women lurks in many dark corners. After transitioning at the age of 40, he remarked: '*Shortly after I changed sex, a faculty member was heard to say: "Ben Barres gave a great seminar today, but then his work is much better than his sister's."*'[7]

Same person, same science, different verdict solely because of the change of gender. This may be only a single example, but what further proof do you need that unconscious (also known as implicit) bias is harboured by some scientists, both men and women?

There is a long history of women, even someone as famous and successful as Marie Skłodowska Curie, being actively excluded from the ranks of science, from the learned societies, and from academic laboratories. It has not simply been a case of women not being educated sufficiently to make the grade as scientists; they have been denied the opportunities until relatively recently to be able to work on (approximately) equal terms with men even when they received that education.

Most children, boys and girls, asked to draw a scientist, still draw a man in a white lab coat, as I'll discuss in more detail in the next chapter. Early on, they imbibe the message that girls don't do the physical sciences, or engineering; or that, by and large, they do not enter the tech industry. Role models who are famous enough to get into the media are few and far between. Wikipedia pages are far commoner for white men from any profession than for women or people of colour. British physicist Jess Wade, 2019 Wikimedian of the year, has made it her project to increase the number of pages specifically for women in STEM so they cannot be forgotten or overlooked.[8] An exemplar such as Marie Skłodowska Curie, usually held up as the role model *par excellence*, is not easy to relate to for a 21st century student. Women's names and faces aren't to be found in textbooks,[9] and female scientists as talking heads on TV are still rare. Yet we need girls entering science more than ever.

I write this book as an academic researcher, a physicist, but there are many paths open to those who are interested in science, ranging from industry to policy, from teaching to design. The world needs production engineers and technicians, designers, and coders. New roles are constantly being created and re-created as innovation changes the nature of work and education. The presence of a diverse workforce in any of these areas is just as important as in research, and so what I write will have much wider relevance than pure academic science conducted in a laboratory. Nevertheless, in what follows, my personal anecdotes will necessarily be based on my time in universities. The evidence about the impact on women will also largely derive from this field, but the nature of the hurdles faced will tend to be the same, whatever path is taken. Impostor syndrome and bias, for instance, are not by any means unique to higher education; nor is harassment. The expectations placed upon women when it comes to 'housework' are different from those placed on men, be it in a professional situation—as with Nüsslein-Volhard's example above—or the more usual domestic sphere, where the Covid-19 pandemic has shone a harsh light on the expectations placed on women as carers across the world of work. This, too, introduces systemic disadvantage.

Progress is being made. The workplace is certainly different from when I first entered it. More women are entering the profession and steadily moving up to positions of seniority, but the speed of this progress is slow, far too slow. Nevertheless, the very fact that bias is now well-recognized and that most organizations are taking steps to eradicate it, means that we are heading in a good direction. Twenty years ago, no one in a lab would have understood what was meant by implicit bias. Now it trips off many a head of department's tongue, although whether they always put into action successful strategies to mitigate it may be a different question. Twenty years ago, people only saw this whole issue as a 'woman's problem' which senior leadership (typically men) did not need to think very much about. Now we hear about men as allies. Men working closely with women to identify the hurdles and find ways to remove them is much more common than it was. If everyone is in it together, we are much more likely to be successful. Many leaders understand the strength of diversity for their disciplines and this will help to move the landscape forward.

If someone is curious, wants to understand how something works (or doesn't work when it's broken) or how we can make the world a better place, science may provide a satisfying answer. Beyond this, for every citizen, understanding science, evidence, and data matters to them on a day-to-day basis: is a vaccination safe? What is the best way of individually contributing to reducing carbon emissions? How reliable are AI algorithms? We are surrounded by the consequences of others' science and innovation, but each of us has our own personal decisions to make on many fronts. We need to be empowered as citizens to make or facilitate such decisions. Unfortunately, school science education too often still comes across as dull, 'mere' facts to be memorized for exams and often without much opportunity for hands-on experimentation or creativity, turning children off science at an early age.

A 2020 report from the Wellcome Foundation, examining the attitudes of teenagers to science education in the UK, highlighted the importance of such hands-on work, stating

> *Practical work emerges as the top motivator for studying science, and students who are traditionally less engaged in science are more likely to want to do more. The decline in practical work from 2016 to 2019, combined with the lack of STEM work placements, is thus a cause for concern and may be contributing to the increase in students who do not view science as relevant to their own lives.*[10]

Lack of time in the classroom, coupled with under-resourcing of school laboratories and increasing concern over health and safety aspects, makes this much harder to provide. Creativity, curiosity, and just plain fun are what science is and should be all about, but too often is not in the classroom, however much teachers would like to make it viable.

Just as I started this chapter with words expressing the thrill I get out of doing science, let me end it with remarks from the Canadian scientist Donna Strickland, only the third woman to win a Nobel Prize in Physics, summing up her own feelings about the buzz that science can give:

> *It is truly an amazing feeling when you know that you have built something that no one else ever has and it actually works. There really is no excitement quite like it — except for maybe getting woken up at 5 in the morning because the Royal Swedish Academy of Sciences and the Nobel Foundation also think it was an exciting moment for the field of laser physics.*[11]

CHAPTER 2

CAN YOU THINK OF A FEMALE SCIENTIST?

Age has not abated my zeal for the emancipation of my sex from the unreasonable prejudice too prevalent in Great Britain against a literary and scientific education for women.

Mary Somerville[1]

I don't think Rosalind saw herself as a crusader or a pioneer. I think she just wanted to be treated as a serious scientist.

Aaron Klug on Rosalind Franklin (quoted by Francis Crick)[2]

A European-wide survey conducted in 2014 showed that a quarter of participants could not name a single female scientist, alive or dead.[3] Not even Marie Skłodowska Curie. Curie may—or may not, as we shall see—be a good exemplar and role model, but the reality is that there have been many women over the centuries who have made significant contributions, and yet are forgotten. In some cases, as Mary Beard has noted in her book *Women and Power: A Manifesto*,[4] women are actively silenced and excluded, in science as much as in any other sphere. Historically, just as often they haven't been given any opportunity to gain the necessary education to enable them to engage. The (UK and Commonwealth) Royal Society didn't admit women to its ranks until 1945, the Chinese Academy of Sciences admitted its first woman in 1955, while the first full member of the French Academy of

Sciences wasn't admitted until 1979. The US National Academy of Sciences was quicker off the mark, admitting its first female member in 1924. Some 20th century female scientists—most notably Lise Meitner, Rosalind Franklin, and Jocelyn Bell Burnell—are remembered as much because they were *not* awarded the Nobel Prize while their male collaborators were, as for what they actually did. There has still (after the 2021 Nobel season) only been a total of four women awarded the Nobel Prize in Physics, seven in Chemistry and twelve in Physiology or Medicine out of a grand total of 342 in those three scientific categories. I've already introduced Nüsslein-Volhard, and we'll meet and hear from a number of these other women throughout this book, as well as more 'ordinary' scientists.

In this chapter I will explore the history behind the absence of women in science, why this occurred and why it matters. I can only present a few vignettes, not a full overview, but there aren't that many well-documented stories to serve as illustration of how women fared working in this sphere from our history. I will identify a few routes by which women made some progress, which will also serve to illustrate the problems they faced. Even when women have broken through, they have often been at pains to stress their domestic virtues. We will see this with mathematician Mary Somerville in 19th century England later in this chapter. Even in the 20th century, Chinese academician and physicist Li Fanghua remarked, '*Whenever I take a female student, I tell her first, "balance your work and family life and never ask your husband to do more housework"*'.[5] Not all modern female scientists would agree with that sentiment, but both of them clearly felt they had to demonstrate they were not neglecting domesticity. In the 21st century we still see vestiges of this expectation when it comes to who takes on the caring responsibilities in a dual career household. The Covid-19 pandemic has highlighted the ongoing unevenness in who gets the time to work in couples. There is a long way to go.

If one goes back many centuries, the name of Hypatia stands out. Living in Alexandria (Egypt) in the 4th century CE, she was a mathematician who taught philosophy and astronomy, and wrote commentaries on existing texts (it is now believed she may have edited Ptolemy's influential *Almagest*) before her murder by a Christian mob. Thereafter there seems little record of women having any prominence in this sphere for more than a thousand years. There was then a brief period, during the 17th and 18th centuries, when women began to make inroads into science, so that there were a few notable firsts. Laura Bassi (1711–78), for instance, the Italian woman who was awarded a doctorate in philosophy from the University of Bologna in 1732 and subsequently became the first woman to be appointed a professor there, a position she held till her death. Even more remarkably, two years before she died, she was appointed to the Chair in Physics, with her husband as a teaching assistant under her. Bassi was, however, not the first woman to receive a doctorate. She was preceded by Elena Lucrezia Cornaro Piscopia (1646–84), who received a doctorate from the University of Padua in 1658. Soon after that, the university's statutes were changed making it impossible for future women to graduate. Two steps forward, one step back.

Aristocracy

Occasionally women made progress not by formal qualifications but by rank. Margaret Cavendish (1623–73) in 1667 is identified as the first woman to attend a meeting at the Royal Society, at that time housed near High Holborn in what is now Gresham College. She was (rather like so many other professional women then and now) regarded as odd, even eccentric, and was known in her lifetime as 'Mad Madge'. Perhaps unsurprisingly, it is her looks and clothes first and foremost that attracted attention when she turned up, as Samuel Pepys (an early Fellow of the Royal Society) describes in his diary:

> *Anon comes the Duchesse, with her women attending her …. The Duchesse hath been a good comely woman; but her dress so antic and her deportment so unordinary, that I do not like her at all, nor did I hear her say anything that was worth hearing, but that she was full of admiration, all admiration.*[6]

These comments by Pepys on Cavendish's dress and appearance, while dismissing her words will, unfortunately, have resonances for many female scientists today.

The Fellows may not have received her well, but she equally did not think much of them. According to Richard Holmes, who has written a vignette about Margaret Cavendish in his book *This Long Pursuit*,[7] she had written arguably the first ever science-fiction novel *The Blazing World*, in which she described:

> *the Royal Society Fellows as various kinds of foolish and predatory animals: 'bird-men', 'fox-men', or 'spider-men'…. criticis[ing] the Fellows' confident reliance on new-fangled optical instruments like the telescope and the microscope.*[8]

She had some reason for criticizing microscopes, having had the opportunity to use them herself. She and her husband, the Marquis (and later, Duke) of Newcastle, had built up an extensive collection of them during their exile as Royalists in Paris in the 1640s.[9]

This hands-on experience meant that she was well aware of the limitations of the lenses of the day, and the way aspects such as illumination could drastically alter what was seen. As she said of their use in *A New Blazing World*: '*for who knows but hereafter there may be many faults discovered of our modern Microscopes which we are not able to perceive at the present*'.

Her negative comments were firmly based in experience, not just derived from being anti-science, as some have characterized her.[10] The maleness of the Fellows also came in for criticism in her writings in both prose and poetry, and she pondered why women should be excluded from the Royal Society, an organization she described as '*dangerous, useless, and deluded*'.[11] It took almost three more centuries before women were first admitted to this august body.

Margaret Cavendish had a strong interest in natural philosophy and cared about the relationship between humans and nature, but she could not have been called a scientist, even had the word existed then.[12] Another woman of nobility who had a much greater claim to being a scientist was Emilie du Châtelet (or Chastellet, 1706–49), who lived in an arrangement amounting to a curious ménage-à-trois with Voltaire (whose real name was François-Marie Arouet), as well as her husband the Marquis Florent-Claude du Chastellet-Lomont. She focussed on the study of mathematics and made a very significant contribution by translating Newton's *Principia* into French, the first such translation, some fifty years after its initial publication in Latin in England, along with a commentary on the work. She published a number of other scientific pamphlets as well. History has tended to see her primarily as Voltaire's mistress, rather than as a scholar in her own right whose translation of Newton has stood the test of time.

She has been described as being trapped between conflicting, unsatisfactory stereotypes: '*the learned eccentric, the flamboyant lover, the devoted mother*', none being exactly complimentary descriptions.[13] For many female scientists ever since, that conflict has persisted, at least between the first and third options.

Helpmeets and Siblings

Emilie du Châtelet was unusual because she had the resources, both time and money, to devote herself to her studies and her work, plus a husband who did not prevent her doing so. Not all husbands were so tolerant: for many women education in the sciences was hard to come by, even with money and time. Science in particular was seen as unladylike, but being a helpmeet was acceptable. Caroline Herschel (1750–1848) started off simply as an assistant to her much-better-known brother William, the astronomer and discoverer of Uranus, polishing mirrors for his ever-larger telescopes. But in due course she was recognized as an important player in her own right. She discovered a series of eight comets in the last years of the 18th century. She was the first woman whose letters were published in one of the Royal Society's own publications *Philosophical Transactions*, although she was never able to set foot in that establishment, despite both her brother and nephew (John Herschel) being Fellows.

Caroline Herschel was, however, tentative about pushing herself forward when it came to science. She had an anomalous position as a sister—not a wife—and a housekeeper to her brother, but with a thirst for astronomy and the ability to do original work. In her own words she says:

When I found that a hand was sometimes wanted when any particular measures were to be made with the lamp micrometer, &c., or a fire to be kept up, or a dish of coffee necessary during a long night's watching, I undertook with pleasure what others might have thought a hardship ....[14]

But as time passed, she felt she needed to put her foot down and focus on her own role if she was to be 'trained for an assistant-astronomer'[15]: she would do her own observations and not simply follow brother William's instructions. When first she saw a (new) comet, she wrote to Alexander Aubert, an amateur astronomer and Fellow of the Royal Society (FRS):

I hope you will excuse the trouble I give you, with my wag [vague] description, which is owing to my being a bad (or what is better) no observer at all ... I beg you Sir, if this comet should not have been seen before to take it under your protection.[16]

Her subsequent letter to Charles Blagden, Secretary of the Royal Society at the time, produced an enthusiastic reply, in which he declared '*I believe the comet has not yet been seen by anyone in England but yourself*'. Furthermore, he immediately communicated the news to colleagues in Paris and Munich and rapid confirmation that the object was indeed a new comet followed. In due course the discovery was formally reported in 'An account of a new comet. In a letter from Miss Caroline Herschel', published in the *Philosophical Transactions*. This was a breakthrough indeed, and Aubert wrote '*You have immortalized your name*'.

Indeed, Caroline's name stood alongside her brother's in astronomy circles at the time, as more astronomical discoveries followed over many years. In turn this led to the radical suggestion that Caroline herself, and not just her brother, should receive a stipend, an idea raised by her brother and communicated through the Royal Society to the monarch, George III. This was indeed an astonishing idea at a time when women could not, for instance, hold a position in a British university, let alone be elected to the Royal Society. Tactfully, this request was formally addressed to the Queen, Queen Charlotte.

> *Perhaps our gracious Queen, by way of encouraging a female astronomer might be enduced [sic] to allow her a small annual bounty, such as 50 or 60 pounds, which would make her easy for life.*[17]

The letter went on to spell out that if this 'bounty' payment were declined, he (William Herschel) would need to employ an assistant:

> *Nor could I have been prevailed upon to mention it now, were it not for her evident use in the observations that are to be made with the 40 foot reflector* [a new telescope], *and the annual increase of the annual expense which, if my Sister were to decline, that office would probably amount to nearly one hundred pounds more for an assistant.*

In other words, the brother was asking for a stipend for his sister in part because she could be cheap labour: half the cost of an equivalent male worker. It is impossible to know whether this is actually what William believed, or if he thought this was a useful tactical approach in order to obtain something, anything, for Caroline (£50 p.a. was actually not a bad rate of pay, for a woman at least, of the day). The tactic worked. Caroline did indeed receive the salary, the first professional salary to be paid to a female scientist in the UK.

Caroline's position was shaken when William married but—whatever her feelings on being the spare woman, the unmarried sister—she did manage to continue her astronomical observations and also continued to be taken seriously by the wider scientific community. The Astronomer Royal of the time, Nevil Maskelyne, corresponded with her regularly. She found seven more comets. So excited was she on discovering the seventh that she rode all the way to Nevil Maskelyne's home, who urged her to write direct to Sir Joseph Banks, President of the Royal Society, a letter that began self-deprecatingly: '*Sir—This is not a letter from an astronomer to the President of the Royal Society announcing a comet, but only a few lines from Caroline Herschel to a friend of her brother's, ….*'[18]

She tried to assert her independence from her brother, moving away to her own establishment, yet inevitably she was always in his shadow and would have felt this. She was clearly hesitant even about allowing herself to believe, or admit to others, that *she* was an astronomer.

Caroline Herschel was able to be an outlier, formidable though her contributions were, by virtue of the circles which she was able to penetrate through her brother. There weren't many women who were able to participate as fully, limited though that participation was compared with that of her brother, in scientific research and circles. Yet her name today is often obscured by the very prominence of her brother as well as her nephew John, who also made significant contributions to astronomical observations. Other women of the same period were afforded few opportunities to move beyond the realm of educators—not necessarily just governesses, although that was often their lot—or studying natural history, a subject considered suitable for women and girls.

Educators

Botany, in particular, was deemed suitable for females. As early as 1741 Madeleine Françoise Basseporte (1701–80) was officially appointed as 'Peintre du Roy, de son Cabinet et du Jardin', in other words, the Royal painter for the Parisian Botanical Gardens; she was also employed to teach painting to Louis XV's daughters. She was sufficiently influential to correspond with many of the key naturalists of the day, including the Swede Carl Linnaeus, famous for his classification system, and more locally George-Louis Leclerc, Comte de Buffon, head of the Jardin du Roy where the King's menagerie was held. However, after her death she seems to have been largely forgotten and her drawings attributed, at least in some cases, to others; misattribution of a woman's work to a man is far from uncommon. One woman making progress was nowhere near sufficient to change the overarching climate in which women found themselves.

In the USA botany became seen as 'suitable' for women, in part due to one particularly influential woman. Almira Hart Lincoln Phelps' (1793–1884) book *Familiar Lectures in Botany* was first published in 1829. Unusually, it was accessible to the general reader rather than being written as a technical tome, which enabled it to penetrate the domestic sphere more widely. Phelps seems to have subscribed to the general (male) view that botany was suitable for a woman both because it was healthy for a woman like her to get outside, and because it enabled her to appreciate God's works in an appropriately lady-like, non-physical way. The connection of women with botanical illustrations seems to have continued through much of the 19th century, but few of them would have been regarded as scientists, using the modern terminology.

Another woman who benefitted from familial connections, but who also fits into the 'educator' category, was Jane Marcet (1769–1858), née Haldimand. Born in London in 1769 into the large family of a Genevan merchant, she married the physician Alexander Haldimand, a political exile also from Geneva who had gained his medical qualifications at Edinburgh University. As he dabbled in chemistry at home, while working during the day at Guy's Hospital in London, she joined him, and when he wrote a pamphlet about this work, she read the proofs for him. This encouraged her to start writing her own books, and she subsequently became a prolific author of elementary texts across science (including botany), as well as religion and economics, referring to the series as *Conversations*. The title referred to its style, not unusual in its day, which consisted of a dialogue between two young girls led by their teacher, who acts to draw out their own thoughts.

The first book covered general scientific themes, such as physics, mechanics, and astronomy, but it was her second book (the first actually published, anonymously, in 1805) which was her most successful. *Conversations on Chemistry, Intended More Especially for the Female Sex* took the work of Humphry Davy, whose lectures at the Royal Institution she was able to attend, and made it accessible to a broad audience. Of course, with a title like that, in her prefaces she had to defend educating girls at all, but she claimed that popular opinion backed her up. As she put it:

> *Without entering therefore into the minute details of practical chemistry, a woman may obtain such a knowledge of the science, as will not only throw an interest on the common occurrences of life, but will enlarge the sphere of her ideas, and render the contemplation of nature a source of delightful instruction.*[19]

This book had a wide readership, including the young Michael Faraday, and was much pirated in the USA. It was many times republished, although Jane Marcet's name remained hidden until the twelfth edition, published in 1832, at which point the fact that a woman had written this popular book finally became acknowledged. It is therefore as an educator, and not as an original researcher, that Marcet should be remembered, but her influence was substantial. She was able to carry out her work and get the books published because she was part of the London literati and science scene and, just as importantly, had a supportive husband. Then, as now, the support of a husband can be vital; their dismay in the face of a woman attempting to crack open doors can be fatal to a woman's ambition and success. I know, from my own experience, just how much difference a husband can make by being supportive and, in my case, taking on a very substantial part of what traditionally is seen as a 'woman's work'.

London's Royal Institution, whose lectures Marcet attended, had opened its doors in 1799, with the aim of introducing new technologies and teaching science to the general public. Davy joined it in 1801. Right from its opening 'ladies' (as long as they were subscribers or 'proprietors') were welcome, with 101 being listed in its first year of existence. They were admitted on the same terms as men, except *'ladies will not be called upon to take any part in the management with the officers of the Institution'.*[20] Davy was seen as a great attraction to the ladies as a lecturer, both because of his good looks and his charismatic style of lecturing. Anecdote says that Albemarle Street in London, where the Royal Institution was located, became the first one-way street in London because it got so clogged up with the carriages of fashionable folk turning up for his lectures. Davy himself was a supporter of the education of women.

Mary Somerville

Until late in the 19th century there wasn't much improvement in openings for women in science. A notable exception was Mary Somerville (1780–1872) who, like Emilie du Châtelet, made her mark via translation, accompanied by a significant and original commentary. In Somerville's case, the translation was from French into English of Pierre Laplace's *Traité de Méchanique Céleste*, with the English title being *The Mechanism of the Heavens*. She was 51 when she published this work; her children were no longer consuming her time and—again and crucially—her (non-scientific) husband was supportive of her role in the scientific sphere. The work was received with great enthusiasm amongst the academic community. John Herschel, a friend of Somerville's family, said of the work '*the same entire absence of anything like female vanity of affectation ... nothing throughout the work introduced to remind us of its coming from a female hand*'.[21] This is praise, but it can be read as faint praise and smacks of Samuel Johnson's comments about women preachers.[22] Herschel goes on in similar vein:

> *We are neither called on for allowances, nor do we find any to make; on the contrary, we know not the geometer in the country who might not reasonably congratulate himself on the execution of such a work.*

However, since Herschel was Somerville's friend, it seems unlikely his intent was patronizing.

Somerville knew Marcet, who was a member of the same élite London coterie. Her writing was hailed with enthusiasm, even if the enthusiasm was tinged with surprise a woman could produce such impressive and original work, so much more than a mere translation. Indeed, her bust can still be found at the Royal Society (although, needless to say, she was not elected a Fellow) for a long time the only visible representation of female scientists. Yet she was clearly aware that not everyone thought a woman ought to be tangling with these matters, writing in her own *Personal Reflections* at the end of her life that '*It has never been my custom to talk of science in my own family circle, far less in society ….*'[23] Additionally, those comments from Herschel raise the question of whether there is a female way to do science, a question whose echoes can still be heard today, even though it is rejected by many women quite explicitly. Interestingly, the first time the word 'scientist' was used in writing to describe those practicing science, was by William Whewell in a review of another of Somerville's books, *On the Connection of the Physical Sciences.*

Science had not yet become a profession. Perhaps that made it easier for Sommerville, since she was not stepping on anyone's professional toes, unlike the women at the end of the 19th century who wanted to enter the medical profession. Like Caroline Herschel, Somerville herself was regarded as sufficiently eminent that she received an annual pension on the civil list, an amount of £200, later raised to £300, comparable to male contemporaries whose names are probably better known as scientists today: George Airy, Michael Faraday, John Dalton, and David Brewster, for instance.

This is not the place to provide mini-biographies of all the (admittedly few) women around the world who made tentative forays into the world of science up to the turn of the 20th century. They were rare and often barely accepted. Herschel and Marcet both had family members who, being involved with science themselves, facilitated their forays into the scientific community, and Somerville's (second) husband was clearly supportive as she rose to prominence. One can see parallels with female authors, however, in the challenges put in their way. Virginia Woolf argued powerfully in *A Room of One's Own* for the importance of private space and financial independence if a woman is to be able to write to her potential and become accepted by her contemporaries. She seemed less concerned about the need to maintain femininity. Jane Austen, it was said by her nephew James Edward Austen-Leigh, was '*careful that her occupation should not be suspected by servants, or visitors, or any persons beyond her own family party*'.[24] That is not too dissimilar to the comment made by Somerville as she started her work, and before she had become famous, '*I hid my papers as soon as the bell announced a visitor, lest anyone should discover my secret*'.[25] She also said, sadly, '*a man can always command his time under the plea of business, a woman is not allowed any such excuse*'. Nevertheless, with the interruptions natural to a Victorian middle-class woman overseeing a house, she had to develop, according to her daughter Martha, '*a singular power of abstraction*'. That is something many a modern scientist would still recognize as a key need. Domesticity remains widely seen as the woman's preserve more than the man's.

In combining the work she did, with maintaining an appropriate (for her times and social class) demeanour, Somerville was obviously successful. A reviewer of her *Personal Recollections* for 'Nature' remarked that:

> *So successfully did she conceal her learning under a delicate feminine exterior, a shy manner, and the practical qualities of an efficient mistress of a household ... that ordinary visitors who sought her as a prodigy came away disappointed that she looked and behaved like any other materfamilias, and talked just like other people.*[26]

One has to wonder at what cost to her 'science' this challenge of hiding her lights under a bushel came. It may well be that her rather sad comment at the end of her life in her memoir *Personal Recollections* that:

> *I had no originality. I have perseverance and intelligence but no genius, that spark from heaven is not granted to the sex, we are of the earth, earthy, whether higher powers may be allotted* [sic] *to us in another state of existence, God knows, original genius in science is hopeless in this.*[27]

is a dispirited nod to the issue. On the other hand, these are words omitted from the second draft of her autobiography, perhaps suggesting she was not entirely convinced by them and was still trying to square the circle of a Victorian woman studying science in her own mind. However, this statement also indicates that she did not expect there to be floods of women scientists, doing original research, following in her footsteps.

Education

Why did I choose these particular women to highlight? In part it's because there are so few women who explored some science whose names are still remembered and about whom much substantive can be written. There aren't that many men's names associated with science after all, and during the 17th, 18th, and 19th centuries, women's education was sadly lacking. Most middle-class or aristocratic women in the UK would have been able to write and read in English. Few would have had any knowledge of Latin, the scientific language of the time. Nevertheless, let me introduce one more woman into this historical tour. This is

Mary Astell (1666–1731). Not a well-known name in the science firmament, although she is appreciated as an early feminist. She came from a modest background, but her uncle Ralph Astell was educated at Cambridge and passed on his love of learning to his niece. When she moved from her hometown of Newcastle to London, she managed to get to know some of the female intellectuals of her day, as well as correspond with some learned men, such as the cleric John Norris. He was so impressed with her theological arguments that he asked her permission to publish the correspondence. She published various other books and pamphlets.

One of these, written in two parts a few years apart, is entitled *A Serious Proposal to the Ladies, Parts I and II. Wherein a Method is offer'd for the Improvement of their Minds* (1694, 1697). It dealt with the issue of women's education. Astell felt that women needed to have a better education so that there were more paths open to them than simply domesticity—running a home for some man, husband, widowed father or unmarried brother being the usual options, or possibly permanently retreating to a religious establishment. It is interesting that she felt so strongly about this, despite herself being both a Royalist and very High Church but she believed—as indeed later, as I've already indicated, did many Victorians, including some scientists—that knowing more about the natural world was simply a way of better appreciating God's works. Despite the future Queen, Princess Anne of Denmark, supposedly initially promising £10,000 to allow such a seminary for young women to be established, Bishop Gilbert Burnet, a leading churchman of the time, felt that this ran the risk of becoming papist in nature and the plan was consequently quashed on his advice. At the time, and for many decades thereafter, it was believed that a well-brought up young lady only needed to know domestic skills, such as how to run a house and embroider, perhaps dabbling in a little sketching or painting, and ideally play some instrument—probably the harpsichord or piano—sufficiently well that she could be called upon to perform in a social setting. You only have to read Jane Austen's novels, for instance, to appreciate how this view was still held in England at the start of the 19th century.

So how does Mary Astell fit into the story of female scientists? She has not previously been placed in such a place, as it is only very recently that any connection between her name and scientific endeavours have been made. Here I must pay tribute to the deputy librarian at Magdalene College, Cambridge, Catherine Sutherland, who introduced me to Mary Astell, having identified a collection of her books in the college's old library, and gave me a wonderful opportunity to explore how Astell had interacted with the published scientific word of the day.[28] As a physicist, such an opportunity is rare and exciting. Many of these books are directly relevant to Astell's own publications, but a couple of them show that this woman, with her limited and unorthodox education, had nevertheless devoted quite a lot of time to studying Nicholas Malebranche's *De la Recherche de la Verité* and René Descartes *Les Principes de la Philosophie*, key texts in the developing field of natural philosophy in the 17th century. With regard to the former, she had gone so far as to compare different editions and translations, attaching handwritten notes at the rear of the second volume of the fourth edition with a pin, which are a copy of a section from the fifth edition of the book.[29]

She heavily annotated her copy of Descartes book, written in French so she obviously had a good grounding in this language too, writing out in her own words (in English), just as a modern student might, what was being argued. Interestingly, looking at the annotations, it is clear that in some places she wrote comments initially in pencil and then inked them over once she'd decided they were correct. Among these annotations, intriguingly, she has not only written her own interpretation of Descartes's words in places, but the blunt words '*This is false* ...' in the margin (Figure 1). I have personally checked where these comments appear. It does indeed appear to be where Descartes was wrong, for instance about the laws of motion, which Newton at the time was approaching in the correct way. It isn't obvious that Astell would have had access to Newton's *Principia* (of course only available in Latin at the time, although it would seem she might have had some familiarity with Latin) so it seems probable her own analysis led her to deciding Descartes' arguments were astray. That strikes me as remarkable, given the isolation in which she must have been working. What other writings she could get hold of, books that she did not personally own, is of course unclear. It seems likely that there will have been other books she did possess (despite a fairly limited financial base from which to purchase such expensive items) which have not been traced. However, an additional annotation in her copy of Descartes demonstrates that she did have access at some level to the journals of The Royal Society, since she explicitly quotes Edmond Halley's paper on evaporation at one point.

These annotations raise substantial questions. First of all, how many other women were actually working hard at getting to grips with key scientific texts of the time? The copy of Descartes that Astell possessed has lain unseen for three centuries; it is pure chance it has come to light, and indeed come to my attention as a scientist, now. We cannot know how many other women's books,

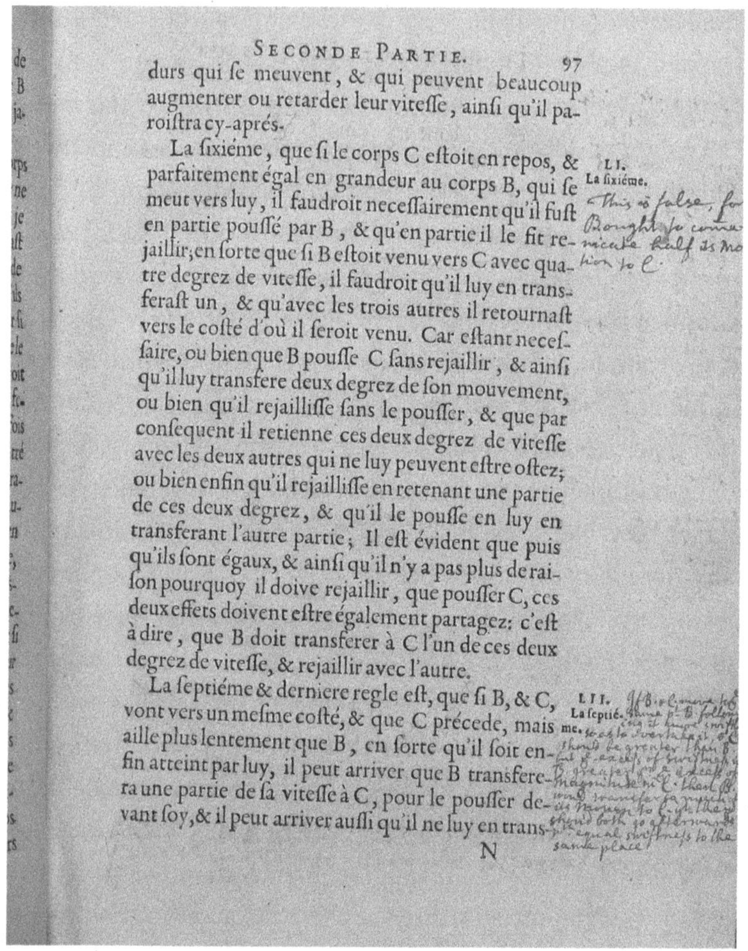

SECONDE PARTIE. 97

durs qui se meuvent, & qui peuvent beaucoup augmenter ou retarder leur vitesse, ainsi qu'il paroistra cy-aprés.

LI. La sixiéme.

La sixiéme, que si le corps C estoit en repos, & parfaitement égal en grandeur au corps B, qui se meut vers luy, il faudroit necessairement qu'il fust en partie poussé par B, & qu'en partie il le fit rejaillir; en sorte que si B estoit venu vers C avec quatre degrez de vitesse, il faudroit qu'il luy en transferast un, & qu'avec les trois autres il retournast vers le costé d'où il seroit venu. Car estant necessaire, ou bien que B pousse C sans rejaillir, & ainsi qu'il luy transfere deux degrez de son mouvement, ou bien qu'il rejaillisse sans le pousser, & que par consequent il retienne ces deux degrez de vitesse avec les deux autres qui ne luy peuvent estre ostez; ou bien enfin qu'il rejaillisse en retenant une partie de ces deux degrez, & qu'il le pousse en luy en transferant l'autre partie; Il est évident que puis qu'ils sont égaux, & ainsi qu'il n'y a pas plus de raison pourquoy il doive rejaillir, que pousser C, ces deux effets doivent estre également partagez: c'est à dire, que B doit transferer à C l'un de ces deux degrez de vitesse, & rejaillir avec l'autre.

LII. La septiéme.

La septiéme & derniere regle est, que si B, & C, vont vers un mesme costé, & que C précede, mais aille plus lentement que B, en sorte qu'il soit enfin atteint par luy, il peut arriver que B transferera une partie de sa vitesse à C, pour le pousser devant soy, & il peut arriver aussi qu'il ne luy en trans-

N

Figure 1 Mary Astell's annotations in her copy of René Descartes *Les Principes de la Philosophie.*
Photo by permission of the Master and Fellows of Magdalene College, Cambridge.

annotated or not, have been tossed away over the intervening years. Secondly, Astell's approach to her studies is notable. As I say, the annotations strike me as akin to a modern student taking a yellow highlighter to a text and spelling out the passage in their own words in the margins. The confidence to disagree with the published word is something that, in my experience, only comes quite late to a research student. Too often a student will read a research paper and feel that their understanding must be amiss if they don't agree with what is there in black and white. It is a clear sign of self-confidence when they are willing to say, 'this has to be wrong'. So, in Astell we have an example of someone who, despite apparent scientific isolation, has nevertheless reached that stage of confidence.

There is one further interesting fact about Astell's life that has been known but little commented on, and that is that she spent five months in 1697/8 working as an assistant to John Flamsteed at the Royal Greenwich Observatory.[30] Flamsteed had 131 assistants during his time as Astronomer Royal, each paying a guinea a week for board and lodging; Astell was the only woman amongst these. Little information seems to be available about what she did during this time, but it says a lot about her (and also about Flamsteed) that she was prepared to undertake such an unusual placement. Her biographer, Ruth Perry, says of her '*Her interest in mathematical concepts is evident from the metaphors she uses*', indicating that her grounding in science and mathematics percolated her wider philosophical writings.[31] In Astell we have, not a woman scientist *per se* (she never published anything on the subject), but a woman who studied contemporary cutting-edge science out of pure interest. How many other such were there at the time, for whom we have no trace?

Long after Astell died, and despite her attempts to improve their lot, women's education in England continued to be narrow and focused on the skills the men folk in the family considered appropriate for future matrimony. (This of course only applies to middle and upper-class women. Most working-class women, just like their brothers, would not have learned to read or write.) To my mind it is hardly surprising that we know so little about women's contributions to science during the past four to five hundred years. Of the ones who are cited, we can see the aristocracy provided one route which might offer an opening; becoming an educator was another; having a family member who was involved in science gave a third route in. And then there was the route that, in du Châtelet and Somerville's cases brought them to wider attention, that of translating. There is a third notable example of a woman who came to public attention due to their work in translating, and that is Ada Lovelace (1815–52). She has become something of an icon for modern day women scientists, and is often claimed, erroneously, as the first person to write a computer program.[32]

Ada Lovelace

Lovelace is very different in many ways from the other women I've described, because she was explicitly given a serious mathematical education. She was the child of the poet George, Lord Byron, albeit her parents separated when she was barely a month old and he never saw his daughter again. Her mother, Annabella Milbanke, was dubbed the Princess of Parallelograms by Byron before their ill-fated marriage, and had been well-tutored in—and shown great aptitude for—mathematics and the other sciences. By the time her daughter was growing up, Annabella was so appalled by her husband's behaviour (incest was hinted at, with his half-sister Caroline) that she wanted to direct the young daughter's attention well away from anything that might have been thought to be poetic or romantic. For this reason, she made sure that young Ada's education was focused on what would then have been called the exact sciences from an early age. Because of the circles that Ada and her family moved in she could get the best tutors and corresponded and socialized with the older Mary Somerville, whom she regarded as something of a mentor. After her marriage to Lord William King (later Earl of Lovelace), whom she had met following an introduction from Somerville's son from her first marriage, Woronzow Greig, she wrote to Somerville

> *I now read Mathematics every day and am occupied in Trigonometry and in preliminaries to Cubic and Biquadratic Equations. So you see that matrimony has by no means lessened my taste for those pursuits, nor my determination to carry them on.*[33]

Grieg also introduced her to Charles Babbage (1791–1871). Babbage had far-sighted ideas about mechanical computational machines (none of which were ever completed in practice), but felt the British scientific establishment were not sufficiently keen or generous in their support. He found what appeared to be a more congenial environment in Turin, where he interacted with Luigi Menabrea (1809–96), who wrote, in French, an account of Babbage's so-called Analytical Engine. Charles Wheatstone (1802–75), another scientist and mutual friend of the Lovelace's, suggested Lovelace translate this paper but incorporating her own additional thoughts. Just like Somerville earlier with Laplace, the translation Lovelace produced in 1843 included, at the suggestion of Charles Babbage,[34] an extensive commentary of her own that took her way beyond the original.[35] And here it was that she explored the idea of what we would now call an algorithm.

Schooling

Lovelace's education was unusual, as I say. The idea that women should be better educated, not least to enable them to become more proficient governesses, was starting to surface in mid-19th century England, however. One of the first serious schools for 'young ladies' was Queen's College in Harley Street, founded in 1843. Here the syllabus included mathematics, something that caused consternation in some quarters as not being suitable. Cheltenham Ladies College was founded 10 years later as the sister school to Cheltenham College for Boys; it also offered mathematics as part of its syllabus aimed at providing 'a sound education for girls'. Before that, schooling would have been much more hit and miss. The kind of school that would have been run—as with many schools for boys—as more a way of disposing of children than actually teaching them much. Charles Dickens' Dotheboys Hall (*Nicholas Nickleby*, 1841) and Charlotte Brontë's Lowood School (*Jane Eyre*, 1847) paint grim pictures of the sorts of education on offer in England around this time.

It isn't that there were no people who had thought about this. Erasmus Darwin (Charles' less famous grandfather) had written a pamphlet in 1794 *A Plan for the Conduct of Female Education in Boarding Schools* about what appropriate schooling for girls might be, to assist his two illegitimate daughters as they were setting up school in Ashbourne in Derbyshire. He had wanted young ladies, not only to learn modern languages (more useful, in his eyes, than Greek or Latin), but also botany, chemistry, and mineralogy. Music and dancing, he believed, were typically 'overdone'.[36] But his forward-looking ideas, like quite a lot of his other proposals, did not find favour as society became more conservative and backed away from the radical ideas that he had promoted and which were associated with the dangers of the French Revolution.

So, it was only in the middle of the 19th century that education for young ladies was seriously considered in England. And even then, science would not have been likely to play a particularly large part of the curriculum. But from the middle of the century on, science started to become professionalized—for men—and women began to get a serious education in growing numbers around the world.

Professionalization of Science

I have dwelt at some length on Mary Somerville earlier in this chapter, because she lived during this period of transition regarding science as a profession. She represents a rare example of a middle-class woman who managed to make some public headway in science to achieve something approaching such a professional life. As the 19th century came to an end, women were starting to make significant contributions in different arenas around the world. They remained a rarity, even if their numbers were rising. They were up against enormous odds: of parents who discouraged them, of lack of financial stability, of universities that would not formally admit them or award them full degrees and of men who actively discouraged them.

Nevertheless, in growing numbers we can find them writing research papers which found their way into professional journals. Examples include Hertha Ayrton, described in more detail below, and the 80 women—'computers'—employed by Edward Pickering at the Harvard Observatory.[37] Most of these were limited to clerical and computing (i.e. data analysis) roles, but Annie

Jump Cannon (1863–1941) has the distinction of being publicly credited, with Pickering, of the Harvard Classification Scheme for stars, unusual in so far as women often don't get the credit they deserve for their research. Kirstine Bjerrum Meyer (1861–1941) was a Danish physicist, the first woman to obtain a PhD in natural sciences in her country and, in 1899, winner of the Danish Scientific Society's gold medal. Netty Maria Stevens (1861–1912) was an American geneticist who discovered sex chromosomes after Mendel's ideas about heredity had resurfaced at the turn of the century. There was slow but sure progress as women managed to get their education and go on to progress professionally.

As science became professionalized, the idea of a PhD being the route to becoming a research scientist gained traction and the availability of PhD training spread, notably from Germany. Whereas mid-Victorian British scientists like Charles Darwin and Thomas Huxley had largely gained their knowledge and skills experientially, in both cases by travelling the world on British ships, that career path was becoming inadequate. Many American universities opened their doors to women undergraduates as, or soon after, they themselves opened. Not so for many of the Ivy League universities such as Harvard, Princeton, and Yale. There, admitting women was much more contested (as described in Nancy Malkiel's 'Keep the Damned Women Out')[38] and did not occur till past the middle of the 20th century, although Cornell did open their doors to women within five years of first opening.

In France the same pattern can be discerned, with most universities being open to women, except Les Grandes Écoles. Indeed, the École Normale Supérieure voted formally to exclude women in 1940, having previously had no such rule (although few women entered), a rule that was only reversed in 1985. During the intervening years women were taught separately at a parallel institution. In Germany, Baden was the first state formally to open all universities to women in 1900, although a handful of women had managed to obtain German PhDs before this date. Notable amongst these was the Russian mathematician, Sophia Kovalevskaya (1850–91), who obtained her PhD from the University of Göttingen in 1874. However, in the 1930s the Nazi regime introduced a cap on the number of women enrolled in universities and the trend in rising numbers was abruptly reversed. In the UK, rather like the USA, most universities admitted women, but in Oxford and Cambridge there was a long delay in granting them the title of their degrees, although they had been able to study in the universities since colleges explicitly for women opened, starting with Girton College (my own *alma mater*) in 1869.

Thus, formal undergraduate and postgraduate education, including in the sciences, was not always accessible for women, let alone easily so. As a result, it could only be via a trickle of individuals that women began to make inroads. One notable woman in UK science, whose travails illustrate the prejudices present at the time, is Hertha Ayrton (née Phoebe Marks, 1854–1923), an early graduate from Girton. A mathematician and a physicist, whilst school teaching for some years she also found time to write up solutions to some mathematical puzzles and invented a 'line divider', which she patented. She continued to study and in due course married her professor from the Finsbury Technical College. Professor William Ayrton was an electrical engineer and the pair of them worked together on various inventions and discoveries (in which there are parallels with Marcet). In 1899, she was the first woman ever to read her own paper at the Institution of Electrical Engineers (IEE) on 'The hissing of the electric arc'; soon afterwards Ayrton was elected the first female member of the IEE. The second, it should be noted, was not elected until 1958.

In 1902 Ayrton was nominated for election to the Royal Society, which required a number of Fellows to sign, approving her nomination.[39] However, the Royal Society of the day deemed her ineligible on the (it would seem hastily trumped up) excuse that she was married: her husband was already an FRS. In the words of their Council: '*We are of the opinion that married women are not eligible as Fellows of the Royal Society. Whether the Charters admit of the election of unmarried women appears to us to be very doubtful*'.[40]

The Royal Society did award her one of their prizes in 1906, the Hughes Medal, in honour of her research on the motion of ripples in sand and water and her work on the electric arc; she wasn't one to work in a narrow silo. She was the first woman to be individually so honoured by them, although Pierre and Marie Curie had jointly been awarded their Davy Medal in 1903. By 1991, when Joan Mason wrote a biography of Ayrton, only nine more women had been awarded any of their prizes.[41] In 2006, when I was awarded the Bakerian Lectureship, I was only the second woman (after Dorothy Hodgkin)[42] to be awarded that particular prize in the physical sciences by the Royal Society.

During the First World War women filled the vacuum left by men going off to fight, and made substantial contributions, but the women concerned are most certainly not household names today. Furthermore, just like women in factories or running companies at the time, once the men returned from the front the women became superfluous again and were sent back to their kitchens and families. For instance, Mary Lowndes is well known within the Arts and Crafts movement for her work with stained glass, but she also was involved with setting up a free training school for female oxyacetylene welders during the war and helping to establish the Women Welders Union.[43] After the war she quickly returned to her artistic work. French-born and bilingual Dorothée Pullinger started work as a draftsperson, but during the First World War she found herself managing around 7000 workers, mainly women, at the Vickers munitions factory in Barry-in-Furness. Some of these women were refugees from France and Belgium, so her language as well as her technical skills were important. She, at least, managed to stay close to engineering, managing Galloway Motors after the war, but most of the munitions' employees had no such luck. A newspaper reported an industrial manager of this time saying: '*the average woman does not possess the same engineering instinct as the average man. For repetition work, yes, but for originality and research, well, there is something lacking ....*'[44] Once the war was over, most women were expected simply to return to a domestic sphere.

Despite attitudes such as this, after the First World War more women did manage to make their mark. Around the world, an increasing number of women were being awarded PhDs and 'allowed'—it's hard to think of any other word—to work in university laboratories. Too often, though, they could not obtain paid employment as faculty members, but had to exist by grace and favour of some (male) head of the laboratory or institute. One woman who found herself in this position was the German mathematician Emmy Noether (1882–1935). She obtained her PhD from the University of Erlangen in 1907 and spent some time there, during which time she was elected to several academic societies and received invitations to speak around Europe. She was invited to Göttingen by the eminent professor of mathematics David

Hilbert to work on some ideas thrown up by Albert Einstein's theory of General Relativity. He had hoped Noether would be officially appointed to a teaching position (Privatdozent), but some of the faculty would have none of it. In his annoyance Hilbert is alleged to have made the unforgettable remark '*I do not see that the sex of a candidate is an argument against her admission as Privatdozent. After all, we are a university and not a bathing establishment*'.[45] It did not cause a change of mind among the faculty. Noether did not obtain a professorship at Göttingen until 1919, despite her great mathematical successes which by then were well recognized and, in 1933, as she was Jewish, she fled Germany to spend her last two years at Bryn Mawr College in the USA.

Between the World Wars

A notable female figure in the interwar years was the Austrian physicist Lise Meitner (1878–1968) remembered, like Rosalind Franklin, as much as for the fact she did not win the Nobel Prize as for what she did. Indeed, for her the case is much clearer that she was actively cheated of a Nobel Prize than for Franklin (who had died by the time her colleagues received the prize), despite being apparently nominated for it nearly fifty times.[46] Cheated probably as much for being Jewish as a woman, the prize was awarded to her long-term collaborator, Otto Hahn, in 1944, who continued to work in Germany after she had had to flee the Nazi regime. Meitner had always had her gender acting against her: when initially collaborating with Hahn she wasn't even allowed into the Friedrich-William Universität's Chemical Institute building in Berlin where Hahn worked. She had even had to be patient simply to obtain the education she wanted: when her formal schooling ended at age fourteen, Austrian universities did not yet admit women. Consequently, she had to wait almost ten years—during which time she undertook a serious programme of self-study—before pursuing her PhD, which she finally obtained in 1905.

Meitner managed to build up a successful career in Berlin, collaborating with Hahn, although for many years she held no formal position. Despite being a high-level participant in many public discussions with the most famous visiting physicists of the day, her brilliance was denied this formal recognition of a faculty post for a long time. As a Jew she then had to face up to the Nazi threat. In the end, she escaped to Norway via Holland only just in time. There, again, she held no formal position and had to watch, after the war, as work for which she had laid the groundwork was attributed solely to Hahn. And he, despite their long friendship and collaboration, kept quiet about her fundamental role. Others simply could not believe a woman could have done the theoretical work underpinning nuclear fission (with her nephew Otto Frisch). This work directly led to the development of the nuclear bomb, to her horror. '*I will have nothing to do with a bomb*', she said when declining to join the Manhattan Project.[47] After the end of the war, in 1946 she went on to say '*You must not blame us scientists for the use which war technicians have put our discoveries*'.[48] Given her horror about how her discoveries were utilized, when she died, Frisch caused her gravestone (she was buried in Hampshire, England) to be inscribed with the words: '*A physicist who never lost her humanity*'.[49]

Between the First and Second World Wars, more women around the world were able to make their mark. In the USA,

Cecilia Payne-Gaposchkin (1900–79) made seminal contributions to astronomy, working at Harvard. Initially Harvard banned women from becoming professors and for years she laboured in low-paid research jobs. It was not until more than 30 years after completing her ground-breaking thesis that she finally became a (full) professor there. German-born Maria Goeppert-Mayer (1906–72) won the Nobel Prize in Physics in 1963 for her work in astronomy but, back in 1930 when she followed her husband to Johns Hopkins University in Baltimore, the university could not imagine employing the wife of a professor and so she was left outside the formal system. Nevertheless, she did keep working, because she enjoyed it so much, feasible as a theorist not requiring laboratory facilities. She eventually became a professor in 1946 when the pair moved to Chicago.

Thus, even the most brilliant women of this period suffered under enormous disadvantages. Universities seemed unable to imagine women as needing to be taken seriously. Barbara McClintock was ignored for years and excluded from the standard career path, but went on to win a Nobel Prize. In her case she worked for many years at Cornell, knowing full well they would never offer her a faculty position. When, in 1983 she won the Nobel Prize, she said:

> *The prize is such an extraordinary honor. It might seem unfair, however, to reward a person for having so much pleasure over the years, asking the maize plant to solve specific problems and then watching its responses.*[50]

She seems to have harboured no resentment towards those who kept her away from the academic ladder. Instead, she showed that excitement and love in her research that is such a hallmark of most successful scientists.

For most of the first half of the 20th century it can therefore be seen that there were numerous explicit and implicit hurdles stopping women from 'making it' as successful scientists. For instance, in my own University of Cambridge, women weren't formally admitted to (undergraduate) degrees until 1948, even if they completed exactly the same course of studies and sat exactly the same exams as the men, although some managed to get PhDs despite this problem. My own department of the Cavendish Laboratory, the Department of Physics, awarded its first PhD to a woman in 1926 to the American Katherine Blodgett, who worked with Ernest Rutherford. (She is probably most familiar to physical scientists for her work on Langmuir-Blodgett films, extremely thin films used to provide anti-reflective coatings on the glass which GE was manufacturing.) Most universities in the Western world were rather quicker off the mark than Cambridge when it came to undergraduate degrees. As we have also seen, Hertha Ayrton was explicitly denied being elected to the Royal Society because she was married; but thereafter a formal exclusion on the grounds of sex was introduced. Although Council reconsidered this in 1922 and 1923, taking legal advice and deciding, based on this, that '*under the Charter and present Statutes of the Society women were eligible for election*'[51] it was not until 1945 when, finally, the first two women were elected to the society.[52]

Even after statutes permitted the admission of women, few made it into national academies around the world for many years. I was startled to discover, upon my election to the Royal Society in 1999, that I was the first female mainstream physicist to be elected. And there have been depressingly few since, although change has recently become significantly more rapid. As we will see later, there are significantly more girls and women entering the biological and biomedical fields than the physical sciences and engineering, and my 'first' is just one example of this, as there had been many female biologists elected by then, some chemists and even a couple of astrophysicists (Margaret Burbidge in 1964 and Carole Jordan in 1990). The various national academies are still playing catch-up; the proportion of women in these remain disappointingly low, although around the world's scientific academies there now seems explicit recognition that there is work for them to do to ensure equality.

Marie Skłodowska Curie

I have, it will be noted, skipped over the most famous female scientist of all. The one most likely to be dredged up from someone's memory when asked if they can name a woman scientist: Marie Skłodowska Curie (1867–1934). Her life is well known and told over and over in simple terms. The struggle of the Polish woman (born Maria Skłodowska) who couldn't complete her education until she had worked as a governess for two years, as agreed with her older sister, to enable the latter to train as a doctor (both of them studied in Paris). The woman who laboured in a cold outbuilding to recover enough radium and polonium to demonstrate their radioactivity, a word she coined. The one who won two Nobel Prizes for her work, the first in Physics followed by a second in Chemistry, but her husband had to speak up for her to be included in 'his' prize (the first one). The one who, after his death, was alternately feted and attacked (for an alleged affair with Paul Langevin, her husband's former student), after which

French society became much more ambivalent about her. As her daughter Eve said in her biography 'Madame Curie':

> *Her origins were basely brought up against her: called in turn a Russian, a German, a Jewess and a Pole, she was 'the foreign woman' who had come to Paris like a usurper to conquer a high position improperly.*[53]

This was also the woman who, during the First World War, developed mobile radiography units to assist wounded soldiers on the Front and constructed hollow needles full of the gas radon to treat infected wounds successfully. The one who did all this while having two daughters, including Irène, who went on to win a Nobel Prize of her own, alongside her husband Frédérick Joliot.

Her life is truly an amazing one, but does this make her a good role model? I think not. The details of her life are, in some senses, so extreme, that she might indeed be regarded as a totally unsuitable role model for young girls and women in the 21st century, yet still, hers is the name that is used to conjure up the idea that if *she* could succeed in science, so could you.

We need to find better role models, more plausible role models. If girls are going to believe that 'science is for them', it is not sufficient to hold up one woman from a period totally different from our own, whose life was one of hardship, if also of fame. Curie was a woman who laboured under extraordinarily difficult conditions and who was internally driven with a great passion so that the hours she worked were extreme, to the detriment of her health. Some scientists will be driven to work as intensely as she did, but it is not necessary to devote one's life so exclusively to the laboratory in order to enjoy science, as I hope to show in subsequent chapters. Nor do I believe that winning a Nobel Prize (let alone two) should be the only hallmark of success. Science, discovery, can be rewarding in and of themselves, without the improbable carrot of a Nobel Prize being dangled in front of one. There are many ways both to enjoy science, and to contribute to the wider community and the overall knowledge base. We need scientists—women and men—in laboratories up and down the land, in hundreds of different roles and disciplines. They aren't simply going to be looking for Eureka moments (which anyhow are infrequent and rarely the way discoveries are made). We need scientists—women and men—in many spheres of life beyond the laboratory, from policy-makers to journalists, from innovators to those in the 'creative industries', of which more later. When girls and young women are being encouraged to think about science as something potentially 'for them', they need to think much more broadly than about wildly successful women as role models, a view many organizations are pushing hard to put across, including in appropriate imagery. In the relatively early years of education, being able to imagine yourself in someone else's shoes, someone who may be only a few years older than you, is much more important than focussing on Marie Skłodowska Curie or other Nobel Prize winners of this world.

Silencing

As I said earlier, Mary Beard has talked about the active silencing of women, identifying the words of Telemachus to his mother Penelope in Homer's Odyssey and amounting to 'shut up', as the first recorded example. In this telling, the son instructs Penelope, who was patiently awaiting the return of her husband Ulysses from the Trojan War, not to speak in public. Were scientifically inclined women similarly actively silenced? Are they still? There are probably few examples of women explicitly being told to shut up in a scientific sphere, but one hardly needs such an instruction for women to feel as if they've been silenced over the decades. Women like Marie Skłodowska Curie, Barbara McClintock, and, much earlier, Caroline Herschel who keep going in the face of active discouragement or non-existent support, are rare. Too many give up.

Take the case of Beatrix Potter (1866–1943), not a woman usually associated with science so much as small furry creatures including her most famous creation, Peter Rabbit, delicately illustrated. Nevertheless, as a young woman she tried to penetrate the scientific establishment, in this case at Kew Botanical Gardens, with her work on fungal spores. She went to see the Director, walking into something of a political minefield there, and she attempted to get her paper published through the Linnean Society (a Society which was quite radical in admitting women as Fellows as early as 1904, but Potter was making her attempt a few years earlier). As her biographer Linda Lear says, '*like other women at the time who attempted to gain a hearing for their scientific research at the Linnean, Beatrix's theories were never seriously considered.*'[54]

So, she was allowed to submit a paper, which was then unceremoniously cast aside. That she was ultimately so discouraged that she gave up her studies should come as no surprise. So, instead of thinking of Potter as a scientist who made important discoveries about fruiting spores, and the dual nature of lichens as a symbiosis of fungi and algae, we think of her, if we recall her at all, as a children's storyteller. It is no surprise that the paintings of animals and plants that adorn her stories are so accurate: she had spent a long time studying natural history.

Silencing comes about in many ways that don't necessarily equate to a literal 'shut up' but have exactly the same effect. Other ways of silencing are more akin to the story of Rosalind Franklin (1920–58), in that their work gets overlooked or credited to others. Franklin oversaw the production of the ground-breaking X-ray photograph that enabled Francis Crick and Jim Watson to solve the structure of DNA, but she didn't get the credit for this at the time. Her work not being appropriately acknowledged, she was silenced in the sense that her scientific voice was not heard in the excitement over the double helix story. The silencing of what she did has reduced, long after her death, as in the last couple of decades her actual science has received increasing, and deserved, attention. Ranging from whole institutions,[55] to a wide variety of single buildings at universities across the UK, and on awards,[56] her name is in frequent use today. (The same is true of Lise Meitner, overlooked in her day but feted today. She has an institute named jointly after her and collaborator Otto Hahn in Berlin,[57] with a building built and named in her honour,[58] also in Berlin, and with numerous awards under her name from different physics organizations across Europe.)

Readers of Watson's memoir *The Double Helix* are left in no doubt about his patronizing attitude towards Franklin as a female scientist, a feeling that she can't really have been more than a skilled technician whose work it was therefore—in the custom of the times—entirely reasonable to poach and appropriate without credit.[59] Early on in his book we are treated to his views on this woman who '*was not unattractive, and might have been quite stunning if she had taken even a mild interest in clothes*'. Because she and her boss at King's College, London, Maurice Wilkins, rubbed each other up the wrong way '*Clearly Rosy* [sic] *had to go or be put in her place*'. One can see this as an (abortive) attempt at silencing.

When Wilkins and Franklin came up to Cambridge to hear a talk by Crick about a putative structure for DNA, things did not improve. Franklin pointed out that the model being presented could not be right, as the amount of water in the model described did not fit the data. Watson admits '*Most annoyingly, her objections were not mere perversions...*'. In other words, she was right. Over lunch, as a result, Crick's mood was, according to Watson, '*no longer a confident master lecturing hapless colonial children who until then had never experienced a first rate intellect*'. This sentence is chilling on many fronts.

Over time, this patronizing attitude faded, and relations between Franklin, Crick and Watson improved. Franklin died tragically young before the Nobel Prize was awarded (and, within the terms of the Prize, she could not have been awarded it posthumously) and Watson, towards the end of his book accepts that much of what was seen as her 'difficult' character actually arose because she was a first-rate scientist who expected to be treated as such but was not. '*She had rebelled against* [King's] *hierarchical character, taking offence because her first rate crystallographic ability was not given formal recognition.*' In the epilogue to *The Double Helix* Watson goes so far as to say:

> *We both* [Crick and Watson] *came to appreciate greatly her personal honesty and generosity, realising years too late the struggles that the intelligent woman faces to be accepted by a scientific world which often regards women as mere diversions from serious thinking.*

The Matilda Effect

Those final words were written in 1968, and yet those struggles *still* apply for many. Added to these, in terms of silencing we can also invoke the Matilda Effect, a term coined by the science historian Margaret Rossiter (1944–) in 1993, whereby a woman scientist's work is attributed to a male colleague. The phrase builds on a phenomenon identified much earlier by the American suffragist and abolitionist Matilda Joslyn Gage (1826–98) in her essay, 'Woman as inventor' (first published as a stand-alone tract in 1870 and later published in a journal in 1883) where she lists many examples of women as the originators of ideas patented in a man's name, often for legal reasons:

> *Nor is woman by law recognized as possessing full right to the use and control of her own powers. In not a single State of the Union is a married woman held to possess a right to her earnings within the family; and in not one-half of them has she a right to their control in business entered upon outside of the household. Should such a woman be successful in obtaining a patent, what then? Would she be free to do as she pleased with it? Not at all. She would hold no right, title, or power over this work of her own brain ....*
>
> *Her husband could take out the patent in his own name, sell her invention for his own sole benefit, give it away if he so chose, or refrain from using it, and for all this she would have no remedy.*[60]

She also cites an anonymous editorial in the *New York Times* considering woman's inventive genius, which stated:

> *The feminine mind is, as a rule, quicker than the masculine mind; takes hints and sees defects which would escape the average man's attention. Women frequently carry the germs of patents in their head, and cause some rude machine to be constructed which serves their purpose. If women would fix their minds on inventions, it is entirely probable that they would distinguish themselves in this line far more than they have done hitherto.*

Another back-handed compliment, perhaps, but indicating that even as long ago as the latter half of the 19th century, women were credited with the ability to be innovative, whether or not this was accorded to their own names or not. Rossiter has continued to push for greater recognition of women's contributions, so that—like Meitner, Franklin, and Bell Burnell—they get due credit, ever since she asked in a graduate seminar in her department of the History of Science '*Were there ever women scientists?*', only to be given the answer '*No. Never. None*'. An answer she said, that was '*delivered quite authoritatively*'.[61] The Matilda Effect continues to this day, as will be apparent when we look later at how women's work is cited in publications. It is in many ways the inverse of the Matthew Effect, also known as the Matthew Principle, coined in 1968 by sociologists Robert Merton and Harriet Zuckerman. I was first introduced to this (in the context of organic chemistry reactions rather than with respect to people) as 'to him that hath shall be given'. In other words, if you already have much, be it of status or wealth, more advantage is likely to accrue to you.

In Franklin's case it was actual appropriation of the data rather than simply (mis)attribution—no one would have believed Crick and Watson could have taken the actual X-ray photographs—but the idea is similar. In Meitner's case it is pretty obvious that Hahn got the credit for ideas that were not initially his, however much he may have developed them and, in his case, been able to publish them widely. Ayrton herself wrote wryly, after Marie Skłodowska Curie's husband Pierre was credited with Marie's work, that

> *Errors are notoriously hard to kill, but an error that ascribes to a man what was actually the work of a woman has more lives than a cat.*[62]

Another example of a woman denied a Nobel Prize, whose contribution was ignored for years, was the Chinese nuclear physicist (working in the USA) Chien-Shiung Wu (1912–97). Two other Chinese-American physicists, Tsung-Dao Lee and Chen Ning Yang, made a prediction about parity in nuclear weak interactions in 1956.[63] They also outlined an experiment that might be able to confirm their prediction. Although it was Wu who successfully carried out that experiment demonstrating parity violation the next year, she was overlooked by the Nobel Committee.

Likewise, the British astrophysicist Jocelyn Bell Burnell (1943–) was the discoverer of pulsars during the course of her PhD but did not receive credit from the Nobel Committee. She wasn't looking for pulsars; in this case, there had been no predictions causing her to seek them out, but their regular signature just jumped out at her when she was looking at radio signals from other sources in space. And, because they jumped out at her, she pressed on to investigate them further. However, her supervisor, Antony Hewish, along with fellow astrophysicist Martin Ryle, got the formal credit. Was she ignored because she was a woman? Or because she was junior and 'merely' collecting data, whereas Hewish had the scientific track record and was the project lead? It's not easy to disentangle.

Bell has been generous in her own summing up about not being awarded the prize. When I interviewed her in 2019, she said that '*she wasn't cross, although a lot of people were cross on her behalf*' at not winning the prize.[64] The logic she gave for this response was that the political animal in her felt pleased that the 'stodgy' Nobel Physics Committee had, for the first time, awarded their prize to astronomy: it meant the door was now open for her community to win further prizes in the future. She was, however, proud that it was 'her' star that had converted the committee to astronomy.

In another interview she said: '*I found that people were much more willing to congratulate me on my engagement to be married than congratulate me on making a major astrophysical discovery*'.[65] One would like to think that that attitude is well in the past but, as we will see later, motherhood—if not simply marriage—still seems to imply to some that a woman has ceased to be serious about her science.

Dorothy Hodgkin, née Crowfoot, is (still) the UK's only scientific woman to win a Nobel Prize, in 1964, and she most certainly did not let pregnancy or motherhood stop her research. A crystallographer, she was one of several women who worked with Desmond Bernal in Cambridge and London during the inter-war years. Bernal himself was certainly a man very supportive of women in science, even if less professional relationships also often developed. In turn, during Hodgkin's long research life, she herself supervised and collaborated with many other women. Crystallography, still, seems to have something much closer to gender balance in its teams than many other branches of physics and chemistry, probably significantly facilitated by the way both Bernal and Hodgkin nurtured female talent passing through their laboratories, including in the latter case, if briefly, Margaret Roberts, better known to the world as Maggie Thatcher.

Hodgkin did not let her gender get in the way of her work, or perhaps even think much about it. When married, but still working under her maiden name of Crowfoot, she presented a key paper at a major meeting at the Royal Society in 1938 when 8 months pregnant. Another long-term collaborator, Nobel Prize winner Max Perutz, referred to her appearance at this meeting in his speech at her Memorial Service: '*Dorothy lectured in that state as if it were the most natural thing in the world, without any pretence of trying to be unconventional, which it certainly was at the time*'.[66]

However, the rest of the world may have been less willing to forget her sex. When she won the Nobel Prize, for her work establishing the structures of the vitamin B12 and penicillin, the press made no attempt to play down her gender. The *Daily*

Telegraph announced: '*British woman wins Nobel Prize—£18,750 prize to mother of three*', while the *Daily Mail* was even briefer in its headline '*Oxford housewife wins Nobel*'. The *Observer*, in its write-up commented '*affable-looking housewife Mrs Hodgkin*' had won the prize '*for a thoroughly unhousewifely skill: the structure of crystals of great chemical interest*'.[67] No male Nobel Prize winner is likely to have their parental status or domestic skills headlined in this way.

The intersectionality of gender and race makes for a double whammy. *Hidden Figures* by Margot Lee Shetterly highlights the massive contributions that black women made to NASA over the years.[68] Most notable of these was Katherine Johnson (1918–2020), whose calculations were invaluable to so much of NASA's work, including John Glenn's ground-breaking orbital mission in 1962. Glenn specifically asked for Johnson to check, by hand, the calculations produced by an electronic calculating machine which he obviously didn't fully trust. Later she was also involved in the calculations underpinning the successful Moon landing of Apollo 11 in 1969. There were numerous other black women helping different aspects of NASA's work, yet for years their contribution was overlooked and forgotten, with NASA's workforce being viewed as primarily white and male. Shetterly's book, later made into a film, brought these women's contributions back into focus to reclaim their achievements and overcome the Matilda Effect.

A recent study has looked at just how damaging disadvantage is for anyone in STEM who deviates from being a white, able-bodied heterosexual man (WAHM).[69] Analysing data for over 25,000 STEM professionals, they considered whether WAHM experienced better treatment and rewards compared with all the other combinations of gender, race, sexual identity and disability status, and demonstrated that WAHM scientists have a significant advantage that act as a premium simply by virtue of their identity, and not down to hard work or other factors that can be worked on to overcome. Their argument is therefore that one should consider mechanisms of privilege as well as disadvantage.

A Female Way of Doing Science?

Perhaps, you are thinking, the women I mention were not silenced, overlooked, expected to stay in the background, or even to stay at home. Perhaps, instead, they just did science all wrong. I have heard arts students (female) complain in recent years when their tutor says something along the lines of 'why can't you write an essay more like a man', with shades of Henry Higgins, as noted in the Cambridge University Student Union Report *Mind the Gap*.[70] I have no idea what that phrase means. Although I have never heard the same complaint equivalently raised regarding the STEM subjects, maybe there are those who think there is a female way of doing science and that the primary activity of a female scientist is to be a woman doing science and not a scientist doing science. However, every scientific woman who is ever quoted on the subject is quite clear: that idea is rubbish.

Take the Canadian laser physicist, Donna Strickland (1959–), who said when she won the Nobel Prize in 2018:

> *But I don't see myself as a woman in science. I see myself as a scientist. I didn't think that would be the big story. I thought the big story would be the science.*[71]

Or the Norwegian neuroscientist May Britt-Moser (1963–), winner of a Nobel Prize in 2014, who expressed the view

> *I often get asked the question about how does it feel to be the first woman to.' Or 'how does it feel to be a leading woman in science . . .' but you know what? I don't think about myself as a woman in these contexts, I think about myself as a human. Thus I feel like a role model for all young people who want to do science—not only girls.*[72]

Or, in a very different sphere, consider the remarks of Marissa Mayer (1975–), former CEO of Yahoo and prior to that part of the senior leadership at Google:

> *I'm not a woman at Google, I'm a geek at Google. If you can find something that you're really passionate about, whether you're a man or a woman comes a lot less into play. Passion is a gender-neutralizing force.*[73]

Women do science as men do science: they do science by bringing their whole selves to the task. Their upbringing may matter in how they think, just as a man's may, or their education or chance encounters they had in the lab, or their mentors and parents. Many things will make the scientist the person they are. We need to throw away the idea that a woman's science is somehow different from a man's simply by virtue of the fact that they don't possess a Y chromosome. We will look in Chapter 5 at recent neuroscience studies which show that, at birth, there is no significant difference between boys' and girls' brains. There are numerous skills a scientist needs, but owning one X and one Y chromosome is not one of them. As far back as 1979, Nobel Prize winner Peter Medawar spelled out in his book *Advice to a Young Scientist*, that there was no basis for a chromosomal difference in scientific ability. As he said: '*The idea that women are, and are to be expected to be, constitutionally different from men in scientific ability is a cozy domestic form of racism*'.[74]

Diversity Matters

Why does any of this matter? I argue that science is done best when it is approached from as many different angles as possible. Following in a straight line, directly on from what has gone before, may lead to new insights but equally may fall into what one might term 'group-think' and be little more than incremental science. Diversity of views—just like diversity in the Board Room—tends to lead to more novelty, creativity and innovation. A US study on PhD recipients and their dissertations across three decades, found demographically underrepresented students innovated at higher rates than majority students, but their novel contributions were discounted and less likely to earn them academic positions.[75] Thus, diversity will lead to better science, not least because a 'white male by default' presumption can lead to some startling omissions in approaches to crucial question. Let me reiterate, this book is primarily concerned with women in science. That is both what there is most evidence about, and what I am most familiar with at first hand. That is not to detract from the additional barriers people of colour—whatever gender—or disabled people face, but I am not well positioned to write about these additional hurdles. They are, however, real and painful and certainly need addressing, as the study I cited earlier spells out in painful detail.[76]

Stanford University's Londa Schiebinger has been articulating this concern for years, and the website she directs, Gendered Innovations, has highlighted some shocking examples.[77] Why, for instance, are the crash dummies routinely used to test car safety based around a 'standard' male? Does women's safety not count (let alone pregnant women, when car crashes are the commonest cause of foetal death associated with injury to the mother)? Training for doctors has historically focussed on how a heart attack presents in a male, although women are known to present differently, leading to higher mortality for women. Drug dosage and safety tends to be evaluated, again, on a standard male whose body mass will, typically, be higher than for the average woman. Hormonal differences may also impact on safety and yet are ignored. Indeed, women's hormones have been regarded as simply complicating matters, instead of being fundamental to their health. During 1997–2000, the US Food and Drug Administration suspended ten prescription drugs producing severe adverse effects on the market, of which eight produced greater health risks in women.[78]

Even the most basic science on individual cells can be fundamentally flawed—and useless—if no thought is given to the sex of the source of the cell. Journal editors are increasingly requiring data to identify cell sex, but until very recently data was simply pooled (or sex was completely ignored), leading to data which in practice may be useless, or worse, actively misleading. This is obviously particularly important in biomedical applications. Caroline Criado Perez has highlighted many of these issues, as well as numerous others in data science, in her recent book *Invisible Women.*[79]

Now of course, it does not require a woman present on a team to spot these potential problems, but it does seem to be the case that for too long the preponderance of men in scientific teams has led to these questions being overlooked. Carol Robinson, a British chemist and the first woman to hold a chair in Chemistry at the University of Cambridge, and subsequently at the University of

Oxford where she was also the first female professor of Chemistry, put it to me like this when I interviewed her at Churchill College:

> *Once we stop talking about what people are wearing, we will have a more inclusive scientific community. And we will have one where everyone can be themselves. I get fed up with people saying 'oh you should be more strident, more aggressive, more assertive', and I think I'm just going to be me … if we all pretend to be more assertive and aggressive then surely we will never move on.*[80]

Diversity in all its forms—not just of gender, of course, but of ethnic origin, socioeconomic background and so on too—matters. Better science will be done if we facilitate diverse teams. So, it isn't simply the moral case, which I would argue is anyhow overwhelming, but there is a business case too for facilitating more women in the scientific workforce. A report prepared for the Royal Society on the business case for diversity examined this in some detail a few years ago[81] and their own commentary concluded, amongst other things:

> *Potential business benefits of diversity can be classified as 'external' and 'internal'. 'External' benefits include reduced costs, improved resourcing of talented personnel, better products and services, and enhanced corporate image; 'internal' benefits are where a greater range of perspectives leads to increased creativity, innovation and problem-solving.*[82]

Enough said. We need more women in our labs, we need greater diversity of thought and insight, and we need to make sure young women are not discouraged from pursuing their dreams. They need to hear, loud and clear, that being 'different' from those they see represented in books and in the media is no obstacle, and that science needs diverse voices and skills.

CHAPTER 3

NOT ALL SCIENTISTS SHOULD BE THE SAME!

Science is not a religion. It's not a set of beliefs that people choose to adopt. Science is people trying to understand the laws of nature. To do that, we need everyone. But till now the talents of women have been underutilized.

Meg Urry (Professor of Physics, Yale)[1]

Scientists are people of very different temperaments doing different things in different ways. Among scientists are collectors, classifiers, and compulsive tidiers-up; many are detectives by temperament and many are explorers; some are artists and others artisans.

Peter Medawar (Nobel Prize winner) in *The Art of the Soluble*[2]

Women, of course, use computers all the time, and my argument is that it's so important that women are part of the industry that's building them too, but we're not making any headway in that.

Wendy Hall (Professor of Computer Science, University of Southampton)[3]

Portrayal of a Scientist

If society is going to attract a diverse range of people to study and work in the STEM subjects, then what a scientist does day by day, and the sorts of people they are, need to be accurately reflected in portrayals, wherever they appear. Every individual should be able to see themselves potentially reflected in the STEM workforce. They need to understand what scientists do and what skills are needed for success. If the general public cannot identify a female scientist by name, then young girls may well feel they have no place in that community when they grow up. What do people think a typical scientist looks like? The image too often conjured up is of a man in a white coat with grey sticking-up hair (the last no doubt based on Einstein), or possibly with a cackling laugh of madness. The reality is of course very different but hard to get through. Search Google or ask children to draw a scientist, these hackneyed and inaccurate images will too often be what is produced.

Fifty years ago, when a test was first carried out to see what children would draw when asked to visualize a scientist, more than 99% drew a man. The percentage who draws a woman has increased over the years: recently around a quarter of children portrayed one when asked to draw a scientist, although these are mainly drawn by girls. All children will, of course, be influenced by the images they see and the messages they receive from the world around them. We need a paradigm shift in portrayals in the media to reflect reality. We need more oft-cited examples of women making a difference. Women such as Italian Fabiola Gianotti, current Director of CERN, or the late US-Iranian mathematician Maryam Mirzakhani who was the first woman to win the Fields Medal (so far of only two), the mathematics equivalent of the Nobel. Portrayals must change at all levels and in all settings if young girls are to believe they belong in the scientific sphere.

Diversity matters in getting the best science done. The moral case is undeniable, but, as I spelled out at the end of the last chapter and as the quotes at the heading of this one show, the requirement to get different viewpoints addressing today's difficult problems, is an equally important argument. In industrial innovation there is the economic case that people who think differently, whose backgrounds are different, will see opportunities and solutions that the boss's clone may miss. Despite often being overlooked, these innovators may be the ones to make the company millions by virtue of these different viewpoints. For all these reasons, diversity is crucial. We need to expand the nature of the STEM workforce if we are to continue to make advances and breakthroughs. (To reiterate, I am using 'science' in this book as shorthand to cover all these elements of STEM, except where the difference between disciplines really matters.) A McKinsey Report from 2018 highlights the importance of how company board diversity by ethnicity and gender improves the financial performance of the company, but the ability to view a problem from different perspectives can also be important at the micro-problem level of a single project, which in turn may positively impact a company's bottom line.[4]

The lone scientist in the ivory tower or home-built lab, as personified by film versions of Dr Frankenstein, is too often the way a scientist is visualized. But the way modern science is done is different, very different, from the days of dilettante gentlemen or aristocrats experimenting at home, and such images are deeply misleading. Men such as the 18th century's Henry Cavendish, the notably eccentric member of the Duke of Devonshire family who worked away in his private laboratory, barely talking to anyone, holding back from presenting his results to scientists at the Royal Society because he was too shy, would not get far today. (Alternatively, this shyness has been interpreted as autism, indicating that neurodiversity is no impediment to scientific excellence either, just another form of diversity.) He converted much of his house into a laboratory, and only communicated with his servants in writing. He was said to be so shy he would '*stand a long time on the landing, evidently wanting courage to open the door and face the people assembled*', according to a biographer in 1851.[5] Nevertheless, while working essentially on his own he discovered hydrogen, the gas he called inflammable air, and measured the mass of the Earth. As we have already seen, Emilie du Châtelet was a woman who also fits into this aristocratic, lone-worker mould, working in her husband's chateau on her translation of Newton's *Principia*. However, not only has science obviously become professionalized into formally constituted laboratories and research institutes, team-working is now more common than not. For many sorts of research, as exemplified at CERN where Gianotti leads overall, but also in many engineering projects, large group projects are the norm.

Team players—a label often applied to women as if it were a negative—are vital. People who enjoy working with others and sharing ideas should be encouraged into the STEM fold, not deterred. We certainly do not need just those who believe in the myth of the lone scientist and act it out, or those who perform a leadership role imperiously, bossing others around instead of working collaboratively. There are those who believe in the domineering way of doing science, but it is not a productive way of working. Large groups of researchers are frequently necessary because so many different skills need to be brought to bear to solve a given topic. Interdisciplinarity, multidisciplinarity, call it what you will, that is the way a significant proportion of science is now done. Judging by the proportion of women who are active in such areas, it would seem they thrive in these situations. We need them.

Take the case of Sarah Gilbert, the Oxford lead on the development of the Astra Zeneca Covid-19 vaccine. As she says in the book *Vaxxers* that she wrote with colleague and co-developer Catherine Green about their work,

> *Scientific discovery on this scale is very rarely a Eureka moment for a lone genius. It definitely was not in this case and we hope we never sound as though we think we did what we did on our own. It was a collaborative effort by an international network of thousands of heroes—dedicated scientists in Oxford and across four continents, but also clinicians, regulators, manufacturers, and the brave volunteer citizens . . . .*[6]

There are so many ways in which to participate in the scientific endeavour which amount to far more than the narrative of the obsessive Marie Skłodowska Curie and her husband beavering away in an unheated cubby hole for years. Each potential scientist will have their own strengths, which might be just what is needed to solve a particular problem. The explicit messaging that Gilbert quotation contains needs to be heard loud and clear.

Examples of prominent female scientists may be relatively rarely seen in the media, but even if you can see a female scientist, how is she portrayed? One of the big events in physics of 2019 was the first imaging of a black hole. Undoubtedly, it was a stunning achievement by a team of researchers about 200 strong, who had been working over several years to make that image possible. Alongside the computer image of the black hole itself, many media outlets ran a picture of one of that large team, a young woman postdoctoral researcher at Harvard called Katie Bouman, looking suitably emotional and excited when the image had first come up on the screen. From that image of Bouman, many different and unhelpful sub-stories and tropes emerged, ranging from an assumption that this was indeed a 'Eureka' moment in science, to the lone genius who did it all by herself, all splattered with some unpleasant misogyny. We need to deconstruct this story.

As Bouman herself wrote on Facebook in response to the way people attacked the photo of her

> *No one algorithm or person made this image. It required the amazing talent of a team of scientists from around the globe and years of hard work to develop the instrument, data processing, imaging methods, and analysis techniques that were necessary to pull off this seemingly impossible feat. It has been truly an honor, and I am so lucky to have had the opportunity to work with you all.*[7]

She was absolutely clear, just as Gilbert has been, this was not the work of a lone genius nor the work of a day; she wanted to pay tribute to the hundreds of key collaborators. These days, science rarely is down to one person or one moment in time, unlike Archimedes and his Eureka moment in his bath. In the case of the black hole visualization, it was a long period of hard slog, not a sudden moment of inspiration that gave rise to the image on the screen.

Team Working

So, let us move away from the idea of sudden moments of enlightenment done by a single scientist. These days, university scientific research is usually done within research groups consisting, typically, of a mix of PhD students and postdoctoral researchers, quite probably also including technicians as well as the team leader, variously called Assistant Professor, Associate Professor, Professor and Endowed Professor in US academic parlance depending on rank; W1, 2, and 3 in Germany; Lecturer, Senior Lecturer, Reader and Professor in the UK and so on, and sometimes just Principal Investigator (PI). For simplicity, I will typically refer to them all as PI or professor for ease, without regard to particular rank. That description of team composition, however, refers to just a single research group and, often, there will be layers of collaboration between research groups on top of the individual group. Research in industry is even more likely to be based around multiple intersecting teams tackling different aspects of a given problem.

In Bouman's example of the imaging of a black hole, a couple of hundred scientists were involved, working with data from eight different telescopes around the world. Using the image of a single scientist was always going to be problematic. While simultaneously giving science a human face, it led to charges that she was hogging the limelight, that it wasn't 'really' her work and so on. No, it wasn't 'really' her work, any more than it would have been if a single man had been photographed in front of the computer screen in the iconic image, but one wonders whether that question would have been asked of a man. The media went further: her role was diminished. According to CNN, she was merely a graduate student. She had indeed started the project as a student, but by the time of the story her PhD was well and truly under her belt and she was about to take up a position as an Assistant Professor at Caltech (the renowned California Institute of Technology), no mean achievement. However, that one image gave ammunition to some internet trolls to imply that Bouman was stealing other people's glory, despite a preprint already submitted for publication with her as a co-author. Women, the message implied, could not have been at the forefront of the discovery, let alone a young and attractive one (although no doubt in part this was why that particular image had been chosen by the media). After the misogyny kicked in, a male collaborator stepped in to speak up for her contribution. Such an intervention may have been needed under the circumstances, but that it was necessary is a sad reflection on attitudes at large.

Most of the science stories that have hit the mainstream news in recent years are likewise driven by large team science. I have already alluded to CERN as an example of such large team working. CERN, based in Geneva, is a European research organization dedicated to large particle physics experiments. It was established in 1954 and operates by running specific experiments with dedicated equipment, all planned over a timescale of many years. Most recently, the Large Hadron Collider was built to explore fundamental questions in particle physics, including looking for the Higgs Boson—which it duly found, roughly where (in energy/mass terms) the particle had been expected. This particle had been predicted as far back as 1964, when there was no way of 'seeing' it with any equipment then available. The plans to build the Large Hadron Collider took around 20 years to come to fruition and another decade to build, before it could be switched on to start collecting data. (It was unfortunate that after only nine days it had to be shut down for another year due to an accident involving a dodgy solder joint.) From the time when the first particle beams were allowed to collide, until the announcement of the 'discovery' of the elusive Higgs Boson was about four years, but to reach that announcement the scale, both in time and effort of the enterprise, was formidable.

A project like this involves a whole range of different people and disciplines. These will include theoreticians and modellers to predict where to look (in energy terms) and how the signature of the Higgs particle will show up; physicists to determine the experiments to be done; engineers to build the equipment; and data scientists and analysts to interpret the outputs from the different recording instruments. But clearly each of these groups cannot operate in isolation. On the contrary, there needs to be dialogue to make sure that what is designed and built is fit for purpose and that people know what is being measured. The importance of good communication skills in STEM cannot be overstated.

From this description, it will be clear that this is not just a case of sharing ideas with someone with a very similar background to yourself but being able to translate your own sphere of action into appropriate words, so that someone from a very different discipline can understand what is being said. Each discipline will have its own vocabulary; indeed every sub-discipline probably will have developed some jargon that 'outsiders' may not appreciate. Communication will also be critical with the politicians and funders needed to approve the construction in the first place, and the public who want to know how the tax-payers' money is being spent. The skill to explain a particular project in ways that are accessible to non-experts (even if scientists from other disciplines, and even more so if they are not scientists at all) has grown in importance over recent years. It is a skill very different from simply standing up and talking to a PowerPoint presentation delivered to your peers. The rewards are significant, but not everyone is necessarily cut out to do this. We need people who are skilled in this sphere, even if it is not a virtue often extolled as scientists start out.

The need for incisive, clear communication about science has never been made clearer than during the Covid-19 epidemic. As governments around the world struggled with appropriate responses to this scourge, they needed to rely on scientists who could communicate complex ideas in ways that they could comprehend, and then politicians had to weigh up this evidence as they made the necessary political and economic decisions. Scientists all around the world who have managed to convey the uncertainties in epidemiological calculations, the relative risk of vaccination for different segments of the population or the reality of hospital wards, have been immensely valuable if not always appreciated by all governments. Communication to general audiences is a skill all scientists should consider honing, whether or not they want that to be their primary focus in their careers. My own experiences highlight some of the associated potential pitfalls.

Talking to the media is a form of communication that many scientists attempt to master, but it is always wise, whether one is male or female, to undertake media training in advance if disaster is not to ensue. This was a lesson I learned through painful experience. At a relatively early stage in my career the group I was involved with was awarded a substantial grant, joint between government and several large blue-chip companies, to work on a class of materials known as colloids. Colloids are not particularly well-known as a class of materials. However, they are ubiquitous in the daily world, turning up in everything from paint to cosmetics, from food to aerosols. They are materials in which one sort of material is finely dispersed in another, with the former existing in particles (although in an aerosol this could be a bubble rather than an actual physical particle) of a micron (a millionth of a metre) or less. Because these particles are so small, they have a lot of surface area associated with them, with a consequent energy cost. In order to reduce this surface energy, the particles tend to clump together if no extra steps are taken to prevent this.

In my naivety, I used the analogy of lumpy custard or a botched white sauce in a press release to try to enable those attending the 'launch' of our new grant to understand what the colloid science was all about. It was downhill from there. Many local radio stations called me up to talk about this and implied—or in some cases more than implied—that cookery was all I, as a woman, was fit for. I was not allowed to speak for long enough to explain the importance of colloid science and that no, I wasn't studying custard itself; that had been merely an exemplar to explain the concepts. The heading in a tabloid that 'Boffins get £3M to study lumpy custard' was indicative of how things went painfully wrong.

It was 15 years after that before I ventured to talk to the media again, during which time I had received some professional media training. Indeed, I had received two different versions of it: one when fronting a Royal Society report on school education on the radio, and one—of a rather different sort—when I won the L'Oreal/UNESCO Laureate for Europe in 2009. Both sessions were extremely useful in their different ways, and certainly gave me more confidence regarding how to steer interviews to cover the points I wanted, rather than simply where the interviewer chose to take me. I am, nevertheless, still a comparative novice. As the Bouman story indicates, there are dangers in fronting or commenting on stories—and not just for women, although they do seem to have a harder time of it, as I will come back to in Chapter 6—but society needs scientists to talk about their work to help them get to grips with some of the big issues of the day and to understand the implications for them as citizens.

Interdisciplinarity

Discovering the Higgs Boson required interdisciplinary working, crucially making sure collaborators in fields far separated were on the same page. I know first-hand some of the challenges of working with colleagues from very different disciplines, from my own research career. To give a specific example, let me illustrate what happened when I was working on starch, that plant-derived material so important in our diet in bread, pasta, and cereals for instance. Starch, found for instance inside ears of wheat, once the bran is removed, is perhaps not an obvious material for a physicist to get to grips with, but I was interested in what determined the way different molecules packed together inside the granules where the starch is stored in plants of different types, and how this packing is disrupted when the starch is cooked.

It is well known that, if you are thickening a sauce or gravy using cornstarch or flour, heat has to be applied for the thickening to occur. We were simulating this cooking by heating starch suspensions in water while looking at changes in the way X-rays were scattered by the molecules in the starch granules, found in the uncooked material, as they swelled and disintegrated. The disintegration and consequent molecular rearrangements are what leads to the thickening desired by the cook. At room temperature we had discovered that, regardless of the source of starch—corn, wheat, pea, barley, and so on—there was a universality in the distance between neighbouring crystalline regions separated by regions of disordered packing. That all the species we looked at had a common repeating distance was a surprise, since it was well known the molecules involved in these different species were far from identical in size and shape. So, I turned to plant biochemists to help out.

The first thing I realized was the need to learn a whole new dictionary of terms. Biology was not my strong point at school. I had given it up as fast as possible, perhaps put off by a rather fierce biology teacher as much as the syllabus, which seemed to require memory work but not understanding, and I had a big deficit in both knowledge and vocabulary. My collaborator, Alison Smith from the John Innes Research Centre in Norwich, was equivalently unfamiliar with the language of X-ray scattering and physics more generally; we needed to work hard to bridge the gaps. Alison and I spent around a year of occasional meetings getting to grips with the approaches of the other and the language we each used. We managed it. We wrote a grant proposal and we were funded (at the second time of asking). We had great fun working together. If we had not got on well together, I am sure we would have given up long before we reached the point of being able to write a grant proposal since, unlike at CERN, it was our free choice whether to put the effort into a joint project. However, by collaborating we were each able to gain greater understanding and insight, and—something that matters of course to the scientist—subsequently able to write some significant papers together.

That ability to get on with someone, to spark off the other, is crucial for a successful collaboration. Paul Nurse, Nobel Prize winner, former President of the Royal Society and Director of the Crick Institute in London, expresses the same view:

> *You can take a horse to water but you cannot make it drink .... you cannot enforce collaboration. For collaboration to work you have to have a common*

> *interest; you also need to have a common chemistry, personal chemistry. You cannot make it happen if either of those are missing.*[8]

Sometimes circumstances mean there may have to be an enforced scientific marriage within a wider programme, but they are less likely to succeed than those where that personal chemistry exists. Or, to put it another way, '*To be complementary of the other … competency, complementarity and confidence, the trick of collaboration is the 3 C's*' as Françoise Barre-Sinoussi, the Director of the Regulation of Retroviral Infections Division and Professor at the Institut Pasteur in Paris puts it.[9]

I mentioned how Bouman had to face misogyny in the wake of the image of her with her computer screen that went viral. For me, talking about my work on starch, which extended in many scientific directions over a decade or more, likewise brought out the misogynist amongst some of my audiences. At one conference, late in the evening in the conference bar, a senior male attendee came up and started chatting to me about the talk I had just given. This was someone definitely ahead of me in the scientific hierarchy, but very quickly his remarks became insulting. He clearly thought it was witty to associate the work I did on the physics of starch with 'domestic science'; he felt it was legitimate to make fun of my work, and of me as a scientist, by demeaning serious physics as 'mere' cookery, implying that that was the only sort of work I would be capable of doing as a woman. As with the 'lumpy custard' story, once again, the fact that the sort of soft matter physics I did was able to be associated with the woman's realm of cookery, made my work an easy target for some males. Mercifully, at this distance in time I forget the full details of his diatribe. Any attempt by me to counter what he was saying, by stressing the excitement and relevance of what I was doing as a serious physicist, was simply talked down. I then did something I can't ever remember doing at any other time, I simply walked out of the bar, feeling most upset by the occurrence.

Luckily for me I was not alone in this conversation. A friend and (now former) colleague, physicist Richard Jones (Professor of Materials Physics and Innovation, University of Manchester), was also present and heard the verbal attack. He had stayed behind in the bar after I left and when I asked him the next morning about what had happened thereafter, it transpired the diatribe had continued, including comments such as *'it must be awful to work with a woman like that'*, with the consequence my friend had also eventually walked out. I find it hard to believe many men will have had to face up to an equivalent sexist attack. I will have much more to say about the way women are pilloried and demeaned in a later chapter, but it's probably worth saying just how much difference it made that I had a witness to the unpleasant challenge. And as an aside to the thrust of this chapter, it is perhaps also worth saying that, when Richard and I reported the incident to the organizers, they acted to bar the senior scientist from ever attending this annual conference again. It has to be said that not all conference organizers are as active in responding to complaints of harassment.

I found that, within about 12 hours of this attack, I was starting to question what had happened. Had it been my fault? Had I done something that provoked the vitriol? Was I to blame? Maybe I was 'guilty' of something intangible so that there was a rational explanation for the vicious remarks. I realized how easy it is to blame yourself for someone else's shortcomings, to assume responsibility and so exonerate the genuinely guilty party, akin to the way in which abused wives and partners respond to much more serious attacks, physical and mental. Putting this to my friend was immensely reassuring. He had seen the way the whole conversation developed from the outset to my departure; he had then himself been at the receiving end of continuing unpleasantness. He could tell me in no uncertain terms I was not to blame, and I mustn't start seeing myself as at fault; the other person was indubitably the guilty party. Had I not had a witness, the damage done to my sense of self would have been much greater. That I still remember this event 25 years later will indicate that it was disturbing enough. For many women undergoing similar attacks, there may be no one to counter the negativity which they then internalize as a measure of their (negative) worth. I hope Bouman's colleagues were able likewise to counter the impact of the trolls so that she was able to move on from the abuse she received.

Such attacks apart, I have always enjoyed the stimulus that working with those from different fields can bring. For me, that has often been the challenge of working with industry, whose endpoint may be very specific and product-driven. Watching paint dry—literally, in a new kind of electron microscope—or understanding why different starches might work better in paint; understanding why some molecular formulations lead to improved consistency in low fat spreads; or working with manufacturers to improve instrument design. All these challenges have had their appeal for me. Not everyone is the same. We need scientists who work in these very different ways to drive our collective knowledge forward and this message needs to be relayed early on so that those setting out appreciate there are different ways of working. There is plenty of scope for people with very different skillsets to make their contribution. Indeed, we don't want or need everyone to be the same, and we most certainly do not need attacks on individuals.

There is evidence to indicate that women are more likely to engage in interdisciplinary research than men[10]; it seems to be something that is of particular interest to them. Interdisciplinary research is, however, not without its own inherent challenges, so engaging in such research may have subtle dangers. It is often talked up by funders, and it is undoubtedly necessary to solve many problems, yet funders too often seem to have problems fairly assessing and funding such work. Within the UK, this problem has been recognized and studied but certainly not resolved. A 2015 report considered what can go wrong in evaluating interdisciplinary work, and how assessment at both individual and team level could be improved.[11] Particular attention was paid to interdisciplinary research within the UK's most recent Research Excellence Framework (REF2021) to try to ensure a level playing field: I chaired the Interdisciplinary Advisory Panel for the exercise, so I know what sorts of efforts were taken, but there will still be suspicion that these will have been inadequate to ensure that those taking on the challenge of working in this way will not have been disadvantaged. If you are a young researcher, feeling unsupported—as many women and other minorities will—it may feel a bridge too far to stray far from a convenient silo. But many major challenges are inherently interdisciplinary, and we need women of talent, just as much as men, to feel confident about entering such areas.

Splitters, Lumpers and Other Types of Scientists

In 1857, Charles Darwin wrote to his friend Joseph Hooker (at the time Assistant Director at Kew Botanical Gardens) that '*It is good to have hair-splitters & lumpers'* a phrase that has stayed in common parlance.[12] Particularly of relevance to classifying living organisms, the phrase distinguishes between those who 'lump' specimens (it was initially used in the context of plants) which are slightly different into a single species, and those who see the smallest differences as indicating the plants must belong to different species. However, the phrase also highlights that it takes all sorts to make the community of scientists. We need diversity, and the range of skills that are required is large: entrepreneurs, policy experts, blue sky thinkers, communicators, and both lumpers and splitters.

In 2014 the UK's Science Council came up with a list of ten different classes of scientists who may operate in different spheres across the scientific workforce, not just the university laboratory, but who are vital for the health of our disciplines.[13] They wanted to counter the image of the white-coated man I refer to at the start of the chapter and listed named examples of those practising in the UK. Their ten classes were: business, communicator, developer, entrepreneur, explorer, investigator, policy, regulator, teacher and technician scientist. Of course, one could—just as with lumpers and splitters—divide up the scientific workforce in different ways, and many people might fit into more than one category. I, for instance, was included in the communicator category but would like to think I might just as well have been thought of as an investigator. Nevertheless, the basic message is the same: to get the best science done, the workforce needs a wide range of different characters, covering a diverse spectrum of skills. And certainly not just men.

Diversity in background is also important. Venkitaraman Ramakrishnan, former President of the Royal Society and Nobel Prize winner (and always known simply as Venki), a scientist born in India but substantially educated in the US who has subsequently spent many years of his research career in the UK, put it this way:

> *I think [being] an outsider allows you to see things with a fresh perspective that isn't obvious to people who are either part of the culture or part of the field. They are used to thinking in a certain way. And you come in there and say, 'Wait a minute. Why is everybody doing it like this?' When people come in from a different field or different country, they bring a new perspective sometimes. That's why I'm a big fan of people moving around in science. Science often benefits from this global [churn].*[14]

However, being different can, perhaps inevitably, come at a cost, and again gender as well as race can lead to an outbreak of hostility when a scientist is placed in the public eye. The experimental evidence confirming the existence of gravitational waves was one of the big media stories in science of the past few years. As in the case of the imaging of the black hole, the public face of the breaking news story in the UK was another woman, Sheila Rowan, former Chief Scientific Advisor for Scotland, and Director of the Institute for Gravitational Research in the University of Glasgow. I am not aware that, as a white woman, she came under attack for fronting the story, but two other women were less fortunate.

When the BBC programme Newsnight ran the story on the evening of the announcement, they (and other media outlets) chose to talk to cosmologist Professor Hiranya Peiris of University College London, a Cambridge-educated woman originally from Sri Lanka. A second expert was paired with her, UCL Honorary Research Associate Maggie Aderin-Pocock, a space scientist and keen science communicator, familiar to UK TV audiences as a co-presenter of the programme, Sky at Night. Aderin-Pocock was born and brought up in London, but her parents are from Nigeria. This pairing of experts caused outrage in some quarters. Both were women, and moreover both were women of colour. The *Daily Mail* ran a story claiming the pair had been selected simply because of their gender and ethnicity because '*Newsnight's Guardian-trained editor, Ian Katz, is keen on diversity*'.[15] UCL vigorously defended the pair, pointing out how well-qualified they both were to engage in this debate, and the *Mail* had to back down. It was an unpleasant episode. However bad the situation is for women, it is worse for women of colour: intersectionality matters. White women almost certainly suffer far less than their black counterparts in the Western world.

This is not to say white men always get off scot-free, although one might argue that sometimes they bring trouble on themselves. I am thinking here of the man who led on another big science news story, about the landing of the *Rosetta* spacecraft's probe on a comet in 2014. Unfortunately, the story became as much about the shirt he wore as about the spectacularly successful space mission which had been carried out with phenomenal precision in order to make the landing achievable. His shirt was, as the *New York Times* described it,

> *a bowling shirt emblazoned with a print of numerous bodacious women in various cleavage-baring poses wearing skintight outfits and toting guns—and bearing an astonishing resemblance to the girls of the just-released* Pirelli *calendar.*[16]

In other words, the shirt was about as unprofessional as it could have been and caused outrage as well as diversion away from the science. The individual subsequently apologized, wearing an entirely subdued and bland hoodie.

The moral of that story is, probably, to think more seriously about appearance. In order to come across as a serious professional, it's probably necessary to be aware of looks as well as words. That is, unfortunately, a trap that usually catches women out far more than men. There have been various social media memes objecting to the way women dress, including those in STEM. There was a meme I briefly came across on Twitter about medical doctors wearing bikinis when on holiday, implying women in bikinis couldn't be serious doctors. Next week, no doubt there will be something different but similar, just as there was a fuss when it transpired that US Congresswoman Alexandria Ocasio-Cortez had once been filmed dancing as a student and this was used in an attempt to denigrate her. Her challenging response was to post a film of her dancing outside her new office in Congress as she took up her position. Women, particularly women of colour, are apparently in the wrong in some people's eyes, whatever they do.

Nevertheless, the question of what to wear when giving a conference presentation, for instance, is one that troubles female students; it has been a question that I have frequently been asked to give advice on. My answer is to wear something in which the student feels comfortable and confident. I don't believe that it doesn't matter, in so far as it is important that it's not (just) the clothing that sticks in the audience's mind. On the other hand, it's probably not a good idea simply to fade into the background either. Over the years I have come to believe that, as I am necessarily a rare female physicist amongst a sea of men, standing out in people's minds can be an advantage. Hence, I have given up wearing sober black, white and grey in favour of bright clothing—but personally I would not be comfortable (nor ever would have been) wearing tight clothing or ultra-low necklines. That is my own choice, but one I am comfortable with.

Likewise, for an early career researcher to look too casual may mean they are not taken as seriously if they wear T-shirt and jeans as if they wear chinos and a shirt with a collar. I think I would apply the same advice to men as well as women, but it probably matters more for women that they do everything that is under their own control to be taken seriously and so reduce the risk of giving rise to misogynistic ire or biased assumptions. Looks and clothing are undoubtedly a double-edged sword for women, however much one might prefer to believe that we are all utterly free to act as we want and to imagine anything else is victim-blaming. Nevertheless, as Nüsslein-Volhard said '*It was very important for me to be taken seriously for my science and not for my looks or other personal accomplishments*'.[17]

Regardless of appearance, what matters to be a successful scientist is how one thinks and acts. Difference is good because of the range of insights that will be brought to bear on a problem. When it comes to invention and innovation, thinking outside the box is crucial and so simply employing people who have the same attributes as everyone else in the room may be unconstructive. Jean Liu, President of Didi Chuxing, the Chinese equivalent of Uber, explained this need: '*In the internet era, the key to a successful business is understanding the customers' expectations—and half the customers are women.*'[18] She has made sure her company is attempting to match this philosophy, with women occupying 40% (admittedly not yet half) of its tech positions in 2020.[19]

Women have been innovating for well over a century, as highlighted by Gage's 1870 essay, discussed earlier. However, because industry typically does not have individual recognizable faces, since so much of what they do is teamwork, it is not always easy to identify 'the' inventor, or find role models in industry to stimulate others. Nevertheless, one name does stand out, who is always credited with the invention of Kevlar, the basic building block of bullet-proof vests credited with saving the lives of many front-line workers around the world. This is Polish-American chemist Stephanie Kwolek (1923–2014), who worked for Du Pont in Wilmington, Delaware. Kevlar (formally poly-paraphenylene terepthalamide) is a type of polymer that is intrinsically stiff at the molecular level, unlike most polymers we use every day. Its discovery opened up a whole area of research, spawning a family of new polymers for use. Industry provides an exciting and satisfactory career for so many, with plenty of scope for creativity.

Liu and Kwolek come from industry and, although as I have said, I can only write this book from the perspective of an academic research scientist, it is important to realize the breadth of other roles open for a scientist to play in modern society. The goal of the Science Council in producing its list of scientist types, referred to earlier, aimed to make this diversity clear; so do my earlier comments about communicators. Many of the issues I describe in this book will apply wherever a person works; they are generic for women scientists. Industry is a major employer, much larger than academia. Not all scientists are either academic or research-focussed; there are many technicians who may be running or building equipment for others, for instance. There are communicators (such as Maggie Aderin-Pocock) and those who run science festivals and museums; their scientific training is fundamental even if not all of them would describe themselves as scientists. These people, with their scientific education, are all playing a crucial role in our society.

Few politicians or policy-makers have scientific backgrounds; Margaret Thatcher in the UK and Angela Merkel in Germany were notable exceptions as world leaders both, as it happens, trained in Chemistry. Merkel's scientific background was particularly noticeable when she engaged with the evidence around the coronavirus pandemic and formed policy based on this evidence, explaining her logic in clear language even when discussing technical matters. In Thatcher's case, her early appreciation of the greenhouse effect and its impact on climate change could be laid at the feet of her scientific training, enabling her to give a powerful speech to the United Nations in 1989, having grasped its importance for the entire world well ahead of most of her peers.[20] More politicians trained in science, when so many of the major issues of the day have STEM issues sitting at their core, could only be beneficial for the public. Climate change, the pandemic and vaccination, AI algorithms or food security are immediate examples which impact on all of us, but the decision-makers are the politicians.

Industry is naturally directly implicated in the solutions to problems such as these, for good or ill. In the UK, the Royal Society recently analysed the scientific and technical workforce, which showed there were roughly 15% more researchers than technicians across the national workforce, although numbers fluctuated year on year.[21] (Technicians are a broad class of key people in a team, such as those who run and maintain equipment, carry out routine tests and so on, as opposed to those driving the original research; they are likely to have fewer, or lower-level, formal qualifications than researchers, but their expertise is often critical for the wider team.) Of the technicians, more than 95% were employed in industry, whereas of the researchers the percentage was significantly lower, at around 80%.

The role of a researcher in industry may be very different from one in academia, although technical roles (other than around teaching) may be more similar in the two spheres. Technicians are almost always going to be working as part of a team, supporting others but often playing a crucial role in innovation and driving a project forward. Academic teams may be quite modest in size, industrial ones are almost always large, with diverse skills being brought to bear on a single problem. Inter-or multi-disciplinarity is the name of the game for them. For many people, solving a specific problem—imagine a goal of making a smartphone lighter, for instance, or a washing machine use less water—provides enormous satisfaction, even if a few months later their focus is necessarily shifted substantially as the next problem comes along. For me, it was that suspicion that I would never be able to follow my nose until all my curiosity was satisfied about a specific problem that put me off moving into industry after my PhD. Other people will react very differently and relish the constantly changing challenges, the satisfaction of solving something very practical and then changing gear as managers move on to the next problem.

Whereas a Henry Cavendish, working in a private house converted to his personal laboratory space, could make significant inroads in his day (and in more than one area), the ability to achieve much on one's own is most unlikely to occur today. Furthermore, when thinking about the problems for which society needs solutions, they tend to be ones that need a multiplicity of approaches. It may be obvious to all that no single individual is going to solve the problem of global warming, to take an extreme example. But one can look at slightly more modest goals and see the same need: how to build an electric airplane, for instance, or how to remediate contaminated soil. It also must be recognized that social scientists must increasingly be involved, another case where interdisciplinary working is crucial. So, if we are to decarbonize the economy, for example, changes in the way we all live our lives are essential. Knowing what the scientific and technical answers are will not resolve many of these issues unless we also get the societal issues right. It is regrettable that society's messaging may imply that 'pure' science is 'better' than applied, so that we see more students wanting to study abstruse mathematical concepts and theoretical abstractions rather than the practicalities of global oceanic currents to understand why the Gulf Stream is in danger of reversing, or how to decarbonize concrete production, a major source of CO_2 emissions.

Finding Your Niche

As an aspiring scientist, working out what your strengths are and finding your niche are not straightforward issues. The routes by which successful scientists fall in love with their subject are hugely variable. It is often only easy to see with hindsight how things clicked into place. Following the advice of Martin Rees, former Astronomer Royal and President of the Royal Society, may be wise but may also be extremely hard to carry out at the time:

> *I think it's very important for a young scientist to work in an area where new things are happening .... because if you go into an area that is fairly stagnant, the only problems you'll be allowed to tackle are those that the older researchers got stuck on. Or couldn't do.*[22]

Equally it may be hard to be sufficiently self-aware, as a young researcher, to know what will work given particular personal characteristics. Nurse describes self-identification in an almost anthropomorphic way:

> *you need to have a feeling for the organism you work with, and I have a feeling for yeast. I can pretend to be a fission yeast and I can do it quite well. I can sort of imagine how a fission yeast feels, and it sounds crazy of course, but I can't so easily imagine how a fly feels.*[23]

Again, it's probably only with hindsight that it's possible to see why something became a good fit, and yet it is so vital to get this right.

Nurse's identification with the organism may seem almost mystical and bizarre, but I can speak similarly about this from personal experience. My PhD and first postdoctoral appointment involved the structure of metals, and the way the constituent crystals were oriented relative to one another in the bulk at so-called grain boundaries (the junctions between crystals), using electron microscopy. The second project, even more than the first, required me to be able to think concretely about the three-dimensional arrangements of atoms. I was not good at this, and the two years of that postdoc appointment were miserable and unproductive. That probably occurred also for reasons far removed from the actual project. These included suffering from culture shock, having moved to Cornell University in the USA from the UK, as well as being a lone female postdoc in an engineering department: I don't think the professor whose group I joined knew quite what to make of me as a female. Nevertheless, although it would have been easy for me to conclude I was useless at research and I certainly had reached the point of deciding I wanted to quit research and become a stay-at-home mum, I had to keep going for entirely non-scientific reasons.

My hoped-for retreat from science was temporarily thwarted (and, as it turned out, permanently thwarted too), since I needed to find another job to allow me to extend my visa while my husband completed his PhD at Cornell. As a consequence, I asked around the department seeking a further position. I ended up joining another professor's group, who needed an electron microscopist to study the failure mechanisms of plastics. I knew nothing about plastics. These materials were not crystalline but amorphous, lacking all regularity in the packing of the polymeric long chain molecules that made up the material. For the first time I did not need to be able to think in three-dimensions and I flourished. Within weeks we were writing our first paper together and I suddenly felt on cloud nine and, finally, discovered a real love of research.

As time went on, I began to think of the molecules I was studying in an anthropomorphic way, rather as Nurse describes his organism in a different context. I could imagine how the long chains of polystyrene would entangle, disentangle and ultimately move past one another (think of snakes moving, or cooked spaghetti falling off a fork) in ways that allowed me to make sense of the way different polymers failed and broke. I had found my niche, but entirely accidentally. I thrived and my career really took off. It was luck. So much is. But I had also demonstrated the ability to pick myself up off the floor after the setbacks of the previous two years. One cannot doubt the importance of both luck and resilience in a scientist's career; finding what excites you is not always easy. Many would-be scientists fail to find the specific topic that would be 'right' for them and so leave the field. Many others, of course, leave because they are discouraged by others rather than by circumstances, something that sadly is much more likely to be the case for women.

The right area for anyone may, as in my case, not be the first one touched. American chemical engineer Frances Arnold, who won the Nobel Prize in Chemistry in 2018, started off working in engineering:

> *When I first studied biochemistry is when I became excited about science. I got a degree in aerospace engineering because I wanted to work on the most complicated things that we could think of, but then when I took biochemistry I realized we can't hold a candle to the beauty and complexity of nature.*[24]

Chance of course comes into play in terms of what opportunities there are available at the time a research topic is chosen. Some people are necessarily steered by external circumstances too, particularly—I believe—those who move into biomedical areas, because they knew someone (perhaps a close family member or friend) who suffered from or died from some specific medical condition. Watching a loved relative suffer over an extended period may provide the motivation to work in such an area. But I believe for many, a general interest in science driven by an overarching curiosity is then impacted by what may be quite random situations and chance, leading to a particular path being chosen. An eminent physicist friend of mine chose his PhD because one potential supervisor he went to discuss a project with smiled at him, sufficient to convince him to join his group. On such a trivial moment a whole successful career was built. I think it is dangerous to assume there is only one topic that will allow a student to flourish, although there may indeed, as in my case, be areas where they will stumble and fail.

Lumpers and splitters; team players and individualists; researchers and technicians, and regardless of ethnicity or gender—all kinds of scientist are needed to ensure the best science is done. We need to remove the idea in our culture that a scientist will always have some particular attribute, be it sticking up hair or extreme shyness, be it a meticulous eye for detail or the determination to stick at a single problem for years on end. Media portrayals could do with rather more variety in the way scientists are incorporated into films or even comedy, as well as in the way science is imagined to be done. Science does not progress by Eureka moments. The love of questioning and the central role of creativity sitting at the heart of a scientist's work is somehow lost in many educational settings. It is to be regretted that the way science is taught at school often does not reflect this. Too often subjects are taught as being divorced, one from another and also as if science is just about dry facts, unrelated to a child's inherent curiosity. I will now turn to consider the social conditioning of early years and how school settings can often deter girls. As will become apparent, this is particularly the case when it comes to careers in the physical sciences, computing and engineering when society still often implies such subjects are not appropriate aspirations for a girl.

CHAPTER 4

WHY EARLY YEARS MATTER

> For whatever reason, I didn't succumb to the stereotype that science wasn't for girls. I got encouragement from my parents. I never ran into a teacher or a counselor who told me that science was for boys. A lot of my friends did.
>
> Sally Ride (first American woman in space)[1]

> There was an assumption that all the girls, even these academic ones, would take domestic science: cookery and needlework.
>
> Jocelyn Bell Burnell[2]

Different education systems around the world obviously have different structures and different issues, but essentially a subject like physics or engineering universally has a higher proportion of men than women studying at an advanced level. What does society do to provoke that difference—and equally, if less studied—why do women form the majority of students studying biology and veterinary sciences? These gender differences in the physical sciences inevitably persist to the top of the profession. Societally, we need to get this right if women are not going to be deterred from a whole range of careers.

Collated data across the national Academies of the Americas[3] in 2015 showed the stark differences by discipline, averaged over the pan-American study. Fellows of a national academy represent the

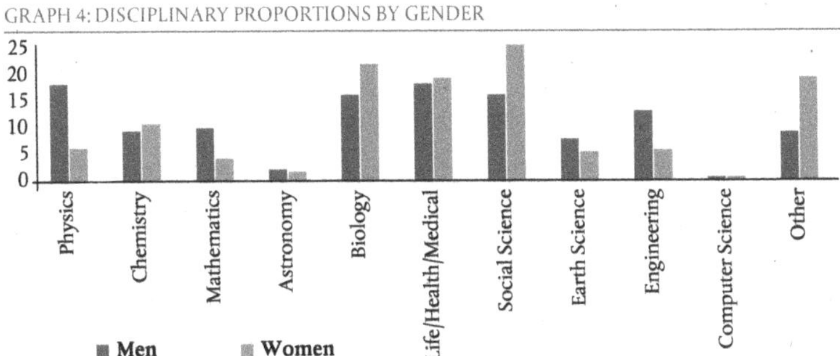

Figure 2 Distribution of men and women between disciplines, aggregated across the Americas' academies.
Frances Henry, *Survey of Women in the Academies of America*, May 2015, https://www.ancefn.org.ar/user/files/SURVEY_OF_WOMEN.pdf. Used with permission.

top echelon of researchers. In physics there are nearly three times as many men as women, in engineering more than twice as many men, but in the life/health/medical category numbers are almost equal, with a slight preponderance of women, and in biology 50% more women than men (see Figure 2).

Stereotypes and the Need to be Smart

Because I am a physicist, and physics tends to show the biggest gender difference of the sciences studied at school, I will concentrate on physics here. And, because physics is crucial to engineering, a lack of girls wanting to do physics at high school and beyond will impact on that discipline too. The absence of women and girls in the subject has been extensively studied in the literature from different viewpoints, testing a wide variety of hypotheses. It would appear that biases against the mind-set that may develop into a practising physicist—or indeed philosopher—start very early. In passing it is worth noting that physics, economics, and philosophy are typically regarded as subjects which require 'brilliance' in ways that subjects such as history or chemistry do not. All three are subjects with a huge imbalance between the sexes. I have always found this rather mysterious but, as the study I describe below suggests, brilliance and 'being smart' are not seen as strengths of girls and women, by themselves just as much as by their male colleagues. Additionally, in subjects like philosophy, for which undergraduate numbers are roughly balanced between the sexes, the number of women dwindle thereafter. As Helen Bebee (a Manchester professor of Philosophy) says:

> *In philosophy, the ability to think with exceptional clarity and rigour about very abstract issues is highly valued, and rightly so; but this is, of course, a stereotypically male virtue. And that means that women have to be that much better than their male counterparts in order to be judged to have the same level of ability. That's just the way stereotypes work: it's much easier to think that someone is an intellectual giant if that's a quality that fits neatly with other things you know about them.*[4]

We will return to explore this sort of stereotypical disadvantage in Chapter 6.

A striking, if relatively small, study led by Andrei Cimpian and published in 2017, considered attitudes amongst five-and six-year-old children in the USA.[5] At five, boys and girls were equally likely to identify both sexes as able to be 'really, really smart', in the phraseology of the paper. However, within a year the girls were less likely to describe themselves that way and, offered different types of games, they tended to stay away from those they thought only the smart kids could handle. Thus, at an incredibly early age, they somehow absorb the idea that boys are smarter than girls, a message that then impacts directly on their choice of activities as evidenced by the choice of games to play that they made. Given it is those adult subjects that are culturally seen as requiring brilliance—or being really, really smart, to use the paper's childish equivalent language—such as physics, philosophy, and economics that are exactly the subjects that the majority of girls and women steer clear of in later life, it would seem these early age influences carry on into adulthood. One could argue that much larger studies would be needed to test whether this is a real effect or not, and whether it applies equally around the world. Nevertheless, it is suggestive of how young girls learn to regard themselves and their gender. Exactly how they receive that message is not made clear in the paper, but many studies have shown how family, school, and the TV programmes they watch, the books they read and the games they play are all feeding into their sub-conscious.

The way girls think about science and scientists is reflected in the 'Draw a Scientist' study, which has been running for more than 50 years in the USA (similar to the European study I refer to at the start of Chapter 2). When boys and girls were asked to draw a scientist, out of 5,000 drawings collected between 1966 and 1977, only 28 were of female scientists, all of which were drawn by girls. A meta-analysis of the decades-long studies shows that there have been shifts, with a significant increase in the percentage of female scientists being drawn by both boys and girls.[6] However, the tendency to draw a woman was much higher at age 6 than 16, showing how stereotyping drives aspiration and expectation and how both these are impacted by societal information absorbed as children grow up.

These studies also align with a slightly different international study from 2018, 'Drawing the Future',[7] asking 7–11-year-olds around the world to draw what they wanted to become and describing how they had heard about that career. The results of this study indicated the familiar stereotyping: boys had a preference for working with things, such as aspiring to be an engineer or a scientist, whereas girls opted for jobs working with people or in the caring professions, with jobs such as teacher, nurse, doctor or vet highlighted. There were also differences between data for developing and developed countries. Children in the former tended to have more practical and highly professional ambitions, such as aiming to be a doctor or a teacher, whereas in developed countries there was a greater tendency to aspire to celebrity culture, including sports or social media. For boys and girls everywhere except China and Australia, either mathematics or science was in the top two most popular school subjects. Most children knew about the career they chose either from a parent or a teacher, indicating, as might be expected, how important these close contacts are in influencing aspirations. This result also shows how important one generation's cultural stereotyping may be, in driving similar stereotypes in the minds of the next. To eradicate such expectations needs explicit and determined effort.

The consequences of the messages children receive from their parents will be reinforced by school teaching. Research in the UK has shown that, although most children up to 10 or 11 years old enjoy and view science in a positive light, less than one fifth of them can imagine aiming at a science career.[8] In the early school years, children like the curiosity-driven exploration that is the basis of all science, but moving on to secondary school (in UK parlance) at 11, with the more rigid curriculum of such schools, turns many off and there is a general loss of enthusiasm for science. By the age of 14, many have completely lost their early interest and become disengaged from science. Consequently, the UK's Institute of Mechanical Engineers recommended that the engineering community should allocate more resources to influence children's interests in this critical 11–14 age group, instead of at a later age once decisions are taken about subject choice for GCSEs (the UK's exams at 16), which will then determine and often limit future choices for study and careers.[9] It is to be expected that other STEM subjects would likewise benefit from a focus on children in this earlier age range, and not be delayed till a time when decisions about career directions are already likely to be fixed. These comments apply to both boys and girls but the problem for girls will be compounded by the gender-stereotypical messages that are all around them.

The Importance of Toy Choices

With parents being so important I will briefly touch on the vexed question of toy choices. Back in the 1970s Lego boxes apparently included the message:

> *To parents: The urge to create is equally strong in all children. Boys and girls. It's imagination that counts. Not skill. You build whatever comes into your head, the way you want it. A bed or a truck. A doll's house or a spaceship. A lot of boys like doll's houses. They're more human than spaceships. A lot of girls prefer spaceships. They're more exciting than doll's houses. The most important thing is to put the right material in their hands and let them create whatever appeals to them.*[10]

At that point it seems clear that Lego, as a company, were well aware of the potential for gendering of toys and did not want to be associated with that. Even more striking, they wanted to advise parents to beware of falling into those self-same traps.

However, at some point, perhaps with an eye to increasing sales, the policy was scrapped in favour of selling different sets to parents of boys and girls. So, pirates suddenly became a boy thing, while in due course girls had their very own series of kits ('Lego Friends') focussing not on adventure but on shops, hairdressing, etc. I understand they did indeed manage to increase their sales to girls, but at what cost to society as a whole? Did girls now think spaceships, castles, adventures, and pirates were beyond their ken? That their role was to worry about shopping and baking cakes?

I found it heartening to see Lego commissioned research to explore how toys impact creativity (not explicitly STEM-related, but definitely relevant to future STEM aspirations).[11] The results, reported in 2021, confirmed that gender stereotypes were alive and well: children, especially boys, preferred creative activities that conform to traditional gender roles, and parents felt limited in the types of creative activities they could encourage their sons to do more than their daughters. Parents still encouraged sons to do sports or STEM activities, while daughters were five times more likely to be offered dance and dressing up than boys. Parents wanted to see more gender-neutral marketing for creative products. On the back of this report, Julia Goldin, the chief product and marketing officer at Lego Group, told the Guardian newspaper that it would cease marketing any of its products 'for girls' or 'for boys'.[12] This definitely seems like a step in the right direction.

Lego is a wonderful product both for imaginative and creative play, but also in enabling children to think and manipulate in three dimensions. Give only boys toys like this to play with, while restricting girls' choices to dolls or tea-sets, and it is perhaps less surprising that girls' mental ability to rotate cubes or read a map, indeed their spatial awareness in general, doesn't improve. However, studies have shown that practice makes perfect.[13] In other words, it is possible to improve a child's ability by more training regardless of gender, which supports the idea that the much-touted idea that men are better than women at spatial awareness may simply derive from what opportunities they've had to practice. (In this particular study I mention, computer games were used.) Without such practice, in later years it is all too easy to say effectively '*see the girls are worse at these tasks. I always knew they couldn't be engineers …*'. It's a self-fulfilling prophecy which recent data on neuroplasticity (discussed in the next chapter) shows up for what it is: not founded on hard evidence. There is also evidence to show that good spatial awareness is important as a predictor of achievement in the STEM subjects so, by excluding significant numbers of children from playing with toys that develop such awareness, they are also being deterred from feeling comfortable with the STEM disciplines later on.[14] Furthermore, if girls subsequently find themselves compared unfavourably with the boys in their class at such activities, it should not surprise us that they walk away from the subjects they appear to struggle with, even if this is nothing to do with their innate talent. By giving all children a wide range of different types of toys we should be able to develop whatever skills they inherently have, however counter-stereotypical. That, needless to say, applies equally to boys in reverse, who may find themselves put off studying subjects like veterinary science or nursing, because they are 'not for boys'. The percentages of boys and men in those disciplines is every bit as low as for physics and engineering when it comes to women. The arguments this way round may get less attention but should not be forgotten.

Toy manufacturers do not always seem to have caught up with these concerns; nor indeed have all toy shop displays. The campaigning group Let Toys be Toys analysed toy advertisements directed towards boys and girls.[15] Boys did not feature in advertisements for baby or fashion dolls; girls were essentially absent from ads for toy vehicles; both sexes featured in ads for board games, art/craft materials, and soft toys. However, the choice of words used in the advertisements involving boys and girls were found to be starkly different. The key words used for boys' ads were 'power', 'adventure', 'control' and, most of all, 'battle'. These are, to use a psychologist's term, agentic words implying actively doing something. Girls, on the other hand, were targeted with passive or merely descriptive words: 'beautiful', 'magic', 'fun', and 'glitter' being the commonest. The most 'doing' sort of word I spotted in that list was 'creative', to go along with 'explore' for the boys. I'll have more to say about creativity in the next chapter: scientists need to be both creative and willing to explore and those words strike me as positive and open-ended. However, I do not believe that a reliance on glitter and magic is going to get a girl anywhere like as far in life as a boy relying on control and power. The message from advertisements is clear: the girl has to sit around having fun and being beautiful, while the boy is meant to get on and do something with their lives, using power and control. That does not bode well for the future. As we will see in a later chapter, the difference in frequency of agentic words used to describe men and women is a significant problem in the professional sphere when it comes to descriptors in letters of reference.

I well remember that when the popular science fiction series Dr Who first aired on television in 1963, with William Hartnell as the first Time Lord, he was accompanied by his granddaughter and two sidekicks, one male and one female. As I recall the main role of the female characters was to scream. I was only 10, but I am sure the impression I gathered that the women were not much use was an accurate assessment of the roles they were expected to play. That, after all, was how the books I learned to read from portrayed things too. Perhaps less screaming, but Jane was always stuck at the bottom of the tree while Peter climbed to the top to rescue the kitten. After that, perhaps Jane was allowed to pass the cakes around. Times have changed and more recently women in the sidekick role have been allowed to do a great deal more, and we have had the first female Dr Who regenerated to delight (at least part of) a new generation, as well as irritate others. But we need to have many more active women on our screens to make sure stereotypes of passive girls and women do not dominate children's minds.

The next toy I want to consider is Barbie, infamous regarding her improbable body proportions, with her unnaturally ultra-small waist and impossibly long legs. Her manufacturers, Mattel, have a curious track record concerning how STEM might relate to their dolls. There was the Barbie doll programmed to say '*Math[s] class is tough*', back in 1992, along with a variety of other phrases such as '*Will we ever have enough clothes?*', '*let's go shopping*' and, more encouragingly '*I'm studying to be a doctor*'. Mattel admitted their error on the mathematics front and the phrase was removed from Barbie's voicebox. However, what had they really learned? In 2010 they produced a Barbie '*I can be a computer engineer*' book. Sounds good. Only it wasn't and in the end was withdrawn with another embarrassing *mea culpa* message from the company. To illustrate why it was quite such a negative book for girls' idea of self-worth and motivation actually to become a computer engineer, here are a few illustrative lines from the text:

> *'I'm designing a game that shows kids how computers work ... You can make a robot puppy do cute tricks by matching up coloured blocks! ... I'm only creating the design ideas ... I'll need Steven and Brian's help to turn it into a real game!'* says Barbie. She then goes on to get a computer virus: *'I tried to send you my designs, but I ended up crashing my laptop.'* So, the boys have to step in *'It will go faster if Brian and I help'*, offers Steven. *'Great!'* says Barbie. The boys conclude: *'Step aside, Barbie. You've broken enough, now.'*[16]

It is hard to see that scenario giving a very positive message for the average girl thinking about future career options in tech. Rather, she will internalize the idea that girls aren't cut out to be coders, computer engineers—or competent at anything much beyond designing something in bright colours involving a cute furry animal. Mattel's formal apology makes clear they recognized they had got it wrong again:

> *We believe girls should be empowered to understand that anything is possible and believe they live in a world without limits. We apologise that this book didn't reflect that belief. All Barbie titles moving forward will be written to inspire girls' imaginations and portray an empowered Barbie character.*[17]

The book was pulled.

In 2021 a new set of Barbies was produced, women with what might be deemed more normal bodily proportions, as well as a variety of skin tones, modelled on real women. This set comprised six women, all from the biomedical arena, including Sarah Gilbert, who led the team developing the Astra Zeneca vaccine against Covid-19. In interviews associated with the launch of these new dolls she expressed her delight at being so honoured adding:

> *I am passionate about inspiring the next generation of girls into STEM careers and hope that children who see my Barbie will realise how vital careers in science are to help the world around us .... My wish is that my doll will show children careers they may not be aware of, like a vaccinologist.*[18]

However, in the book *Vaxxers* she co-authored with Catherine Green, she is quoted by the latter as saying 'something like': '*This is 2020. Why are we discussing women scientists? I'm not a woman scientist, I'm a scientist and more than half my colleagues are women and we do the job.*'[19]

The frustration at being seen as a woman first and a scientist second is palpable. She is not alone in this frustration, as will no doubt be clear from quotes throughout this book.

What Interventions Work?

To turn specifically to physics, where the evidence is strong internationally that girls stay away, a whole range of interventions have been tried in different studies, not all of which have had any discernible impact. In the UK, the Institute of Physics (IOP), the relevant professional body, has worked hard with schools to understand better what is going on and what might reduce the gender gap in those who study physics at A level, that is for the 16–18 age group. One of the more shocking findings, in my view, was a study which showed just how much the type of school attended made a difference in the numbers of girls pursuing the subject at this higher level.[20] Girls-only schools saw about twice as many girls proceed to later years' study of physics as coeducational establishments. This certainly gives the lie to the idea that 'girls just don't like physics', although, when I was interviewed on the radio alongside a (female) headteacher about the study, that is precisely what she tried to argue. It suggests that some combination of peer pressure and teachers' messaging and expectations leads to the discrepancy, although other factors (such as parental backgrounds of those who send their daughters to single sex schools, or that such schools are able to attract better qualified teachers, since teaching quality can be key to pupil's interests) may also be relevant. The IOP's continuing work has shown just how much the whole school ethos matters in the way children make choices and, specifically, the likelihood that girls may opt for physics.[21]

A range of studies, often quite small but coming from different age groups and geographies, all confirm that boys get more attention in the classroom.[22,23] Some of that attention may be because the boys are (stereotypically) less well behaved, less inclined to sit still and be quiet than the girls. But whether the teacher is telling a boy to shut up or praising them for getting a sum right, the fact is they are the focus of more of the teacher's time. Is that likely to encourage the girls? The IOP, as part of its focus on increasing the diversity of students studying physics, has produced a list of top ten tips that teachers should enact, of which one highlights the dangers of not noticing how boys and girls interact differently with them. As one of these recommendations, it states:

> *You might be surprised at the ratio of different genders asking or answering questions in your class. Keep a note yourself or ask a colleague or student to observe one of your lessons and keep count.*[24]

If teachers do not pay attention to girls, whether or not they get good grades (and their grades these days, at least in the UK, are often better than boys) they may not feel good about themselves.

More specifically, girls may fail to see images, in textbooks or on TV, which look like plausible role models for them, to help them imagine a future for themselves in STEM, although this is changing: a search on Google for photographs of scientists produces far more images of women now than a few years back. Additionally, if the only women they ever hear associated with science are the long dead and impossibly over-achieving Marie Skłodowska Curie or the Nobel-deprived Rosalind Franklin, who died young, then I doubt they are likely to feel it is a career for the likes of them. Nevertheless, this quote from an interview I did with Melanie Welham, Executive Chair of BBSRC (one of the main UK science funders) at Churchill College, sums up the frustration many women feel about stereotypes and how early they start:

> *Can we move on from the stereotypes? When my daughters were in primary school, only 7 or 8, I went in to do some experiments with them .... and there were a couple of boys who were really curious, they sort of looked at me ... and came up to me and said 'are you really a professor' because I clearly didn't fit the mould for whatever a professor would look like.*[25]

The evidence shows that girls who get A grades at GCSEs in the STEM subjects are much less likely to take them at A level than boys with equivalent grades. It isn't ability that's stopping them. It is a lack of feeling they fit in, that society would countenance them in those disciplines. If teachers make not-so-subtle remarks discouraging them from the sciences—*'do you really want to do physics, that's unusual for a girl'*, or even the phrase I've heard quoted by an irritated mother of a daughter, *'you do maths like a boy'*, whatever that might mean—then how many 16 year olds have the courage of their convictions to pursue their interests against such messaging? Some do, but we have no idea how many are deterred along the way. The primatologist Jane Goodall said of her own growing up

> *My mother always taught us that if people don't agree with you, the important thing is to listen to them. But if you've listened to them carefully and you still think that you're right, then you must have the courage of your convictions.*[26]

She persisted; others may not.

Again, I should stress, the converse problems may apply to teachers' attitudes towards boys.[27] Many parents may have encountered this sort of simple-minded sexism themselves at parents' evenings when discussing their children. I once was told by an English teacher (a woman) straight out that 'boys can't do English', thereby apparently consigning 50% of the population to the dustbin of literary endeavour. Looking at any list of best-selling authors would seem to contradict this teacher's assertions; there are plenty of extremely successful male writers and always have been. But with attitudes like this around it is not surprising that nearly three quarters of the A level English cohort over the recent past were girls; boys are discouraged, although probably usually in more subtle ways than this particular teacher's upfront remarks.

School ethos matters, and the evidence suggests that coeducational schools provide a rather different environment from single sex ones. Gather a group of practising adult female scientists and ask them if they went to single sex schools, something I have done including with international audiences, and it is noticeable, not to say depressing, how many have done so. I certainly did. How many girls in coeducational schools, remembering that most schools are mixed, are lost because of their experiences with peers and teachers in such a school? I may have been the only girl in my own, admittedly small, 'science sixth form' who wanted to study physics at university, but no one attempted to tell me I should not, let alone that I could not because it wasn't suitable for a girl. In fact, it was almost literally the day before I went to Cambridge University as an undergraduate that my apparent oddity was first brought home to me. A family friend, a biology teacher at (the then all male) Harrow School, was having a house party. My sister and I were invited along, presumably to dilute the all-male atmosphere. I well remember the shock I received in attempting to socialize with these young gentlemen, with conversations which went along the following lines (I paraphrase):

BOY: *'Hello little girl, what do you do?'*

ME: *'I'm going to Cambridge tomorrow to read Physics.'* Boy recoils in horror. End of conversation. I was not a social success that evening.

That was the moment when it first dawned on me that I was not regarded as 'normal'; that my aspiration was considered unusual. It did not, however, shake my belief that studying physics was what I wanted to do, or make me feel that *I ought* to be doing something else.

The nature of science, and especially physics teaching itself, has also been scrutinized around the world. A Luxembourg study of 14-year-olds led by Sylvie Kerger set out to test the premise that science is considered to be inherently masculine and hence being interested in science may threaten how girls view themselves and their femininity.[28] They set out to test this by altering the way the topics were taught, concluding that girls' scientific interest was higher when concepts were presented in the context of feminine topics, and boys when presented as masculine topics. My own approach, were I a school-teacher, would be to give plenty of examples and context when teaching, so that everyone can relate to the lesson regardless of gender or background. The press release I was sent when the article was published, instead, distinctly worried me, stating that

> *when scientific concepts in physics, information technology, and statistics were presented in a female friendly way—as for example relating to online shopping or cosmetic surgery—the mean level of girls' interest rose. However, the boys' interest in these topics simultaneously decreased.*

It is a fine line between reinforcing stereotypes and keeping children interested.

Another study, this time from the UK, considered the 'girly girl', someone who was hyperfeminized.[29] It was presumed that such a girl would be more prone to influence by their peers, indeed a bit 'vacuous' as it was put in the paper. One teacher was quoted as saying '*Oh yeah you can usually tell the girls that want to do Physics, they look a bit tomboy-ish*', and a (male) student said:

> *When I picture someone who's really feminine they just … they're someone who isn't that focussed on doing something when they're older. And say something like a physicist—it requires independence and wanting to be it.*

Stereotypical comments like these abound in the wide-ranging quotes given in the paper. The authors conclude that their findings added to other evidence that the '*domain of science is constructed as, and perpetuates, a masculine epistemology which is excluding of the feminine (and therefore often to girls, who are demanded to engage with femininity)*'.

In other words, the more our society propagates stereotypes, the harder it is for girls to feel comfortable entering the hard sciences. That statement will be true whatever their talent or background.

Indeed, the existence of stereotypes which girls may wish to buck compounds the problem. The ideal, or even simply the idea, that one could be interested in make-up as well as physics may just seem too hard, pushing the STEM subjects beyond a girl's imagination or aspiration. An American social sciences study led by Diane Betz showed that, girls who were struggling in their early teens with STEM subjects, became even more disaffected when shown a role model who appeared able to combine femininity and success in STEM.[30] So, are role models a good idea?

Role Models and Images

'You can't be what you can't see' is an often-quoted mantra, implying that role models are necessarily helpful but, as this last study suggests, this may not always be the case. However, there is some evidence to suggest it can help in some situations. Another study from the US, where decisions about which discipline to major in are left relatively late in university teaching terms, indicated that there is a positive effect on whether or not to choose a particular STEM major if the student has a higher proportion of classes led by a faculty member similar to them.[31]

A significant study of results from the international 2015 Programme for International Students Assessment (PISA) showed how role models—in this case mothers—impacted on girls' interest in science in countries around the world.[32] Their results showed that maternal influence was strong, but mediated by national characteristics, as evidenced by the Harvard Implicit Association test as a measure of that country's tendency towards gender-science stereotyping, along with the Human Development Index (HDI) and the Gender Gap Index (GGI) as assessments of socioeconomic development and gender equality. The higher the level of gender equality, the greater the gender gap in STEM graduates, a fact that was assigned—albeit indirectly—to the idea that gender equal countries tend to have high welfare safety nets.[33] In other words, in less gender equal countries, more girls saw STEM careers as well-paid and providing a better future for them than in countries for which the fear of being left unsupported was less acute. This to some extent is consistent with both boys and girls being more likely to aspire to a 'celebrity' career in more developed countries (as indicated in the Drawing the Future study discussed before), even if this is something statistically very unlikely to transpire, while solid well-paying careers were more highly regarded in less advantaged countries. It has been highlighted that, even in a country widely perceived as one of the best for gender equality, namely Sweden, the gender gap in STEM subjects persists at high school and beyond.[34] The availability of options seems, in itself, to be insufficient to overcome cultural messaging and occupation segregation by gender.

From this brief consideration of recent international studies exploring the reasons girls choose not to pursue the STEM subjects, particularly the non-biological ones, it is clear the reasons are many and varied, and therefore so must interventions be. Different young women will be affected differently by their social and familial environment and messaging, as well as their own innate talents. The messaging they receive can be internalized in very different ways. For instance, a girl told 'girls don't do maths' or 'girls aren't good at maths'—as regrettably so many still are, by both peers and teachers—may either take this message at face value and so turn her back on any subject that relies on mathematics; or see this as a challenge and become determined to prove the nay-sayer wrong. Feminizing the teaching by choice of contextual example may be just the thing to excite one girl and cause another to walk away in disgust. There are no easy solutions until culturally we remove these stereotypes once and for all. This might also facilitate more boys considering careers in biological and medical sciences, as well as in the caring professions.

Everyone imbibes stereotyping messaging, and its impact can be surprisingly strong. First studied by the American, Claude Steele, in the context of black students doing worse at university than their prior grades indicated they should have done, so-called stereotype threat has become a major topic of study. Steele's book *Whistling Vivaldi* covers many instances when negative stereotypes provoke inhibiting doubts and high-pressure anxieties in the mind of an individual when taking a test, for instance.[35] Thus, as one specific example amongst many others, if girls are consistently told that girls don't and can't do mathematics or science, this will lodge in their sub-conscious to the extent that it will cause anxiety during tests, so lowering their performance. This effect then becomes self-reinforcing: if they do badly in one test they may do badly in the next because they 'know' that first result arose because they are a girl. It becomes a self-fulfilling prophecy leading them to walk away from the subject.

It has been reported that simply being reminded about the association between gender and maths ability, perhaps by being asked to state one's gender before doing a maths test, can provoke stereotype threat.[36] Under such conditions, women perform worse on challenging (but not easier) maths tests, not because of innate ability differences but because of concerns they might confirm the negative stereotyping of women being bad at maths. Women using a fictitious name, thereby unlinking their true self from the maths test, showed significantly higher performance, and reported less self-threat compared with those who used their real names.[37] Men were unaffected by doing this.

There is an extensive literature on stereotype threat as related to girls and women in the STEM disciplines and what interventions may help to overcome the issues raised. For instance, it has been shown that asking women to write about their values, so-called 'self-affirmation' reminding them of the positives in their life, helps them to overcome the disassociation they may feel between being female and the stereotypical male STEM world.[38] This is certainly a low-cost intervention but, again, may only work for some people. For me personally, if I had been told to write a paragraph about things in my life that I valued before a physics test, I would probably have found it such an irritating distraction that it seems unlikely it would have improved my performance. Amanda Diekman and her colleagues identify a series of interventions that may help in different situations and different institutions, to produce a more inclusive environment whereby stereotype threat based on gender (and ethnicity) may be reduced in the STEM disciplines.[39]

As their article discusses at some length, this turning away of girls from subjects they believe they are bad at might be countered by teachers encouraging a 'growth mindset'. This is a concept originating with Carol Dweck, who developed this idea towards the end of the last century, and discussed it comprehensively in her 2006 book, *Mindset*.[40] As she put it:

> *Individuals who believe their talents can be developed (through hard work, good strategies, and input from others) have a growth mindset. They tend to achieve more than those with a more fixed mindset (those who believe their talents are innate gifts). This is because they worry less about looking smart and they put more energy into learning.*[41]

It is not the case that boys are inherently smarter and girls will never achieve excellence in these disciplines. Encouraging girls to recognize this fact will enable more of them to succeed. However, as long as the message is retained that having flunked one mathematics test, for instance, they are doomed to flunk the next, with peers and teachers compounding that message, girls may continue to turn their backs on the STEM disciplines.

I think it is clear there is no simple fix to facilitate the entry into higher study of all those girls who, possibly privately and invisibly because of the fear of appearing unfeminine, enjoy physics and allied subjects *and* who would thrive in pursuing the subject. Societies around the world, particularly in the developed countries, seem largely stuck with a mentality which cannot shift from the view that boys think and do while girls are meant to nurture and care. Yet, as I have already stressed in the previous chapter, the evidence is that innovation and discovery would do better if we had a broad range of perspectives brought to bear on any problem. We also need a creative imagination to venture into totally new territory.

CHAPTER 5

CREATIVITY IS NOT JUST FOR ARTISTS

Sciences provide an understanding of a universal experience. Arts are a universal understanding of personal experience. The arts and sciences are avatars of human creativity.

Mae Jemison,[1] First African American woman astronaut in space

Art and physics are much closer than you would think. Art is based on mathematical principles like proportion and harmony in the case of music. At the same time, physics is based on symmetry principles, so beauty is at the foundation of nature.

Fabiola Gianotti[2]

Always think is there another way of looking about this problem. … creativity is often at the edges, boundaries between disciplines, or subject areas. It's putting things together that you often don't put together …. If you want to be creative, explore the edges.

Paul Nurse[3]

There are those who seem to believe that science is not creative. There are those who seem to think that women scientists, in particular, cannot be creative but should be left to the drudge work of cataloguing and collecting data. When I was setting out as a young researcher, I remember my father expressing this opinion to me, without apparently recognizing how much he was writing off half of humanity as only worthy of being drones—or spotting the message he (most certainly not a scientist) was giving to me, at an early stage in my career.

Although there seems very little hard data on differences in 'creativity' between the genders, the idea that men are more scientifically creative does seem to permeate people's thinking. It's something that many people 'know' without looking any further. I would hazard a guess it's because most of the discoveries that people associate with scientific creativity have been done by men, because, as we've seen, there weren't many women allowed in the lab. So, although it may be true that the names of the greats in science, whose discoveries everyone has heard of ranging from Newton to Einstein, from Darwin to Crick and Watson, are all male, statistically it was bound to be so.

What Does Neuroscience Tell Us about Brain Differences?

If you believe the work of UK psychologist Simon Baron-Cohen, there is something distinctly paradoxical about such a presumption, that it is the males who do the whizz bang stuff, leaving the women to grind away at data collection. He has come up with the idea of two kinds of brains, so-called systematizing and empathizing, described in his book *The Essential Difference*.[4] His theories have grown out of his work on autism, something which has been (at least until recently) more associated with males than females: he identifies autism as the 'extreme male brain'. Systemizing he defines as '*the drive to analyze or build a rule-based system*' and he sees this as commoner (but of course not uniquely so) in men than women. Women, on the other hand, are thought to be more likely to be empathizing. By this he means an '*ability to recognize another person's mental state ("cognitive empathy") and the drive to respond to it with an appropriate emotion ("affective empathy").*'[5]

Although he does not go so far as to say all men are systemizers and all women empathizers, his work—in which he talks about the brain being hard-wired differently in men and women—has often been criticized along these lines. Baron-Cohen and I were paired in a debate organized by a Cambridge student body some years ago, under the challenging title 'Why women can't do science'. He clearly recognized he couldn't accept this premise, despite all his written words and papers and began by saying that of course he accepted that women could do science. It made the debate less interesting for the audience. Nevertheless, he stuck to his ideas that the tendency for males to want to take things to pieces, or to catalogue 'things' as opposed to worrying about how people might feel, was innate, conferred by biology in the hormones the foetus is exposed to and resulting from evolutionary trends by which the male is the hunter, whereas the female reproduces and nurtures. More recently, he has tended to soften this position, for instance, in his latest book *The Pattern Seekers*.[6]

Gina Rippon refers to work such as this as 'neurosexism'. Her book, *The Gendered Brain*,[7] looks at recent neuroscience evidence for, and even more against, the idea of hard-wired differences between men and women. There have been a range of easily accessible texts discussing the limitations of such views. Cordelia Fine in her books *Delusions of Gender*[8] and *Testosterone Rex*,[9] and Angela Saini in *Inferior*[10] point out numerous failings in such a deterministic view of the brain. The 'evidence' from functional neuroimaging, using functional MRI (fMRI) to study brain differences, has been shown to be nothing like as clear-cut as some studies have claimed. As Fine has stated elsewhere, such

> *research was found to support the influence of false-positive claims of sex differences in the brain, to enable the proliferation of untested, stereotype-consistent functional interpretations, and to pay insufficient attention to the potential plasticity of sex differences in both brain and mind.*[11]

Part of the early problems stemmed from the small number of individuals involved in each study, so that the statistical significance of any differences seen was often uncertain and the analysis may have been chosen to highlight differences rather than similarities between male and female brains.[12] Modern studies are now able to deal with much larger cohorts, however, which should reduce such errors and huge brain imaging data bases are available for analysis (such as UK BioBank). Nevertheless, there remains a strong focus on any differences perceived, rather than similarities, leading to over-interpretation of small differences.[13,14]

Modern science has shown just how plastic the brain is, whatever its state may be at the moment of birth.[15,16] Babies' and children's brains rapidly reconfigure as they are exposed to new information, through sight, touch, smell, and hearing. If children are exposed to different sensations, the way their brain develops will alter. A study of Romanian infants who were essentially deprived of human contact during their early years and months, because of the way the government operated orphanages at the time, shows just how that deprivation can impact on normal brain development and behaviour.[17] In healthy children, brought up in more normal circumstances, interaction with parents and teachers will systematically impact on the newly formed—and also pruned—connections in the brain as they grow up. Their brains are anything but hard-wired and immutable.[18]

This process of adding and pruning connections goes on throughout life, even if the rate in an adult is significantly reduced, as highlighted by a much-cited study of London taxi drivers (first published in 2000[19] and followed up later in 2009).[20] Eleanor Maguire and her team from University College, London, showed that a particular region of the brain—the posterior hippocampus, a region associated with spatial awareness and representation—was significantly increased in size in the taxi drivers, who need to spend years learning the detailed map of London, relative to those of control subjects who have not done so. Furthermore, some years after retirement from taxi work, this increase in size had vanished. The brain is constantly remodelling itself, which is one of the reasons patients hit hard by a stroke or major brain injury may slowly begin to regain functions lost due to the trauma. It is also why, as discussed in the last chapter, if you give children three-dimensional puzzles to work with, they can improve their ability to visualize three-dimensional problems. The authors also suggest that their work can be extended to aid spatial learning in the general population.[21]

To the 19th century (male by default) scientist it was probably 'obvious' that the fact that the average male's brain was larger than the average female's, 'proved' that the male was more intelligent than the female. The mathematician and statistician Alice Lee was the first to disprove this presumed correlation between cranial capacity and intelligence in her doctoral thesis, results subsequently published in one of the Royal Society's own journals.[22] It is fittingly ironic that it was a woman whose research first destroyed this link, since it meant women could not be lumped together as less intelligent than men simply because their brains (and skulls) were on average smaller.

So at birth, or at any point thereafter, the size of the brain cannot be used to indicate anything about intelligence and whatever the newborn brings into this world will be modified day by day. Thus if, as is done in much modern neuroscience research, an fMRI is taken at some later date, what is observed will be the product both of any innate differences and all that has happened thereafter. Even so, such images show only the most dubiously distinct differences between men and women. Attempts to demonstrate that the connectivity between the left and right hemispheres is much greater in women than men, allegedly related to the empathizing tendency as well as to greater verbal fluency, or that the parts of the brain associated with verbal tasks differ in men and women, show only marginal effects. The trouble is, interpretation of fMRI is non-trivial, with relatively poor spatial resolution: it can only distinguish quite large regions, and most certainly not at the level of individual synapses (connections). What is actually being measured is the oxygen level as a proxy for brain activity, and there is a great deal of data analysis required to produce the pretty-coloured images with which we are now so familiar, and which so often pepper popular science programmes. Because this is all labour intensive and, as anyone who has ever had an MRI scan will know, fairly unpleasant for the participant, usually the numbers of participants involved in such studies are quite small, meaning results may be confounded by statistical noise in the small numbers, rendering solid conclusions hard to draw.

Unfortunately, scientific journals, where academics wish to publish their results, do not respond well to reports of null or statistically insignificant results. This is a major problem across all science, since null results may be extremely important. Only when authors can at least claim some sort of difference is a publication likely to ensue and then their press office may hype subtle differences up as 'stark' or 'significant', when in reality they are small and quite likely not, in a statistical sense, significant at all.[23] As Rippon and colleagues express it '*dimorphism, the existence of two distinct forms, is not an accurate way to characterize sex/gender differences in neural phenotype*'.[24] In other words, there are overlapping characteristics between men and women and the evidence is not robust that how adults perform is simply related to how much testosterone they were exposed to *in utero*. Some males may indeed be more aggressive, or more prone to systematize and less verbally fluent than some women. But it is not hard to find women who are aggressive and men who are not. Generalizations and stereotyping, even if they had any statistically valid basis, are not helpful for any of us, yet this is what is so naively done time and time again.[25] We are all individuals and should be treated as such.

In the creativity saga, a different element that is typically associated with invention and innovation is risk-taking, another trait more commonly associated with men. In order to come up with a new hypothesis, it is rarely sufficient just to build incrementally on existing data. Paul Nurse was quite explicit about this '*I took a risk at the beginning to work on something that wasn't that interesting for most people*'.[26] He was referring to yeast, not a fashionable organism when he started working on it. His career grew from this first step. But risk-taking is not enough; you also need to be fascinated by what you do when you take these risks, as he went on to explain: '*As a young scientist, you should identify problems that are interesting to you, take a risk and stick with them. Let your passion lead you*'.

Is It a Myth that Men Take More Risks than Women?

The evidence does seem to support the idea that men are more willing to take risks, including intellectual risks, but not in all circumstances.[27] Even gamblers may be risk-averse in some situations and how willing an individual is to take part in a specific activity may depend on circumstances,[28] such as whether others are watching.[29] This difference in behaviour is a trait that adolescents are particularly likely to exhibit, behaving very differently on their own than when in the company of their peers, as Sarah-Jayne Blakemore explains in depth in her book *Inventing Ourselves*.[30] So, it is too simplistic to say that A is more likely to take risks than B, because this is not an absolute. Nevertheless, when it comes to intellectual risks, there is evidence to suggest that males are more willing to take risks than females; the study by Byrnes et al. show this as being an area of risk where one of the larger gender differences was found in a meta-analysis.[31] However, I'd introduce four caveats here.

Firstly, what may be true of the average man and the average woman in an average situation is not a good guide to the behaviour of any particular individual in a specific situation. Just because, on average, women may be less inclined to dream up a brand-new hypothesis on the back of a couple of outlier points on a graph, does not mean some women will not be perfectly capable and willing to do so. Secondly, even one particular individual may respond differently in different situations. As Venki described in his memoir, *Gene Machine*: '*A germ of an idea was slowly forming in my mind but seemed so risky that I wanted to be secure enough in my personal situation to have the guts to tackle it*'.[32]

There's nothing like job security and a decent salary to encourage being daring: even future Nobel Prize winners aren't always going to be willing to go out on a limb without thought for the consequences. Thirdly, none of this actually says anything about innate characteristics so much as the way we bring up our children, encouraging boys to dare and girls to play safe. If societally we treated boys and girls equally, who knows what the 'average' girl might be prepared to risk. All these points suggest that it doesn't make much sense to say girls are unlikely to be creative because they aren't willing to take risks. Nevertheless, as Giaonotti spells out clearly, science does involve significant risk-taking, whatever your gender:

> *Research is all about looking for the unknown, and going beyond what we understand. Part of this endeavour means taking risks. Investing your time in something where you do not know what the result will be is an essential component of scientific research.*[33]

What Do You Mean by 'Creative'?

My fourth point is more subtle but opens up a huge can of worms. Many people, when they use the word creative, actually mean something very different from science. They are referring to art, poetry, and novels, composing music, and producing TV dramas. But these are exactly the areas that many girls are pushed towards at school, away from subjects such as sciences. Why should this be so, if men are more creative than women? This is not a topic that has received much attention.

There are a lot of confusing arguments getting mixed up in some of the ways the word 'creative' is used and abused. Firstly, we have the longstanding view in some quarters that science is inherently not creative; I will expand on this misconception below. Then we have the idea that STEM is a boy's subject and by encouraging women into this arena we are implying that the women's work in the arts subjects is not as valuable as men's. I have already discussed the way education does, or rather does not, encourage girls into STEM at some length. Finally, both those points encompass two tropes that I think are extremely damaging: that Arts and Sciences need to be pitted against each other in some weird sort of zero-sum game and that stereotypes are all we need to understand what is going on. There is a lot to unpick here to understand how all these feed into messages about the sciences, and whether girls 'belong' there, which I think contributes substantially to why there are relatively few girls and women to be found in many of the STEM subjects in a range of different societies.[34]

Too often, both historically and currently, science is seen as antithetical to curiosity and exploration, as the opposite of creativity. When William Blake famously carved '*Art is the tree of life. Science is the tree of death*' in 1826–7 on the side of his engraving of Laocoön, closely followed by Thomas Carlyle's damning 1833 statement that '*The Progress of Science … is to destroy wonder, and in its stead substitute Mensuration and Numeration*', this presumption that science is not creative was put firmly into words. For many the world can still be viewed with the Arts and Sciences in explicit opposition, with science being seen as sterile and boring to many. Historically, however, it was probably easier to be a polymath, crossing the divide, as illustrated by two individuals from the early 19th century: Humphry Davy and Erasmus Darwin. These two intellectual leaders could happily write both poetry and, in the former case, lead the Royal Institution in London and, in the latter, be asked to be the King's Physician (he declined) and be a lynchpin of the Lunar Society (which also involved James Watt and Joseph Priestley, amongst others).

Carlyle's statement implied that by analysing some physical phenomenon, such as a rainbow, the wonder was lost. For me, as a scientist, that interpretation is baffling: the wonder is as much in the understanding as in the observing. Clearly not so for Carlyle. In this vein also sits the poet John Keats who allegedly said, of Isaac Newton, although probably with facetious intent, that he *'has destroyed all the poetry of the rainbow, by reducing it to the prismatic colours'*,[35] before proposing a toast to him. A similar sentiment was expressed in his poem *Lamia*. Underlying these various sentences is the mistaken belief that scientists cannot be creative, a view reinforced in some, but by no means all, quarters that the 'creative industries' are somehow divorced from science and technology, relating to literature, art or music.

One website describes the creative industries as those *'businesses with creativity at their heart—for example design, music, publishing, architecture, film and video, crafts, visual arts, fashion, TV and radio, advertising, literature, computer games and the performing arts'*.[36] No mention of science there. This is bizarre, both because you cannot do good science without being creative and, equally, many of the creative industries have science at their core. Video games, for instance, would be a non-starter in the market without a heavy input from technologists. For realism, the very way images are constructed relies on the laws of physics; illumination or shadows won't look convincing if you don't recognize that light travels in straight lines and doesn't go round corners (crudely speaking; you can create situations where it does in an optical fibre, and that requires an amazing act of creativity to dream up, but that is an unnecessary level of detail that doesn't spoil my basic point). The more artistic designers could not produce such creative products without the science input; trying to put a boundary fence around different disciplines is a nonsense. It is also no way to permit the economy to grow through diverse approaches to innovation. A fit-for-purpose education system has to allow students to see the world in the round, even if ultimately they specialize in the specific area they love. Of course, a definition such as the one above for the creative industries is just that: a definition; but it implies a way of thinking that is misleading and which definitely does mislead many. Other definitions of the creative industries are much more broad-minded and pay appropriate attention to the role of technology in design, architecture, and gaming.

My favourite description of the creative process in the mind of a scientist is due to Peter Medawar, who declared in his 1969 Romanes Lecture in Oxford:

> *All ideas of scientific understanding, at every level, begin with a speculative adventure, an imaginative preconception of what might be true—a preconception that always, and necessarily, goes a little way (sometimes a long way) beyond anything which we have logical or factual authority to believe in.*[37]

A speculative adventure feels just right for how to set out into some new uncharted territory where the unknown lurks, hopefully just ripe for picking.

How Hard Is the Boundary between Arts and Sciences?

Pitting the Arts against the Sciences is a dangerous game—be it when regarding government funding or pushing boys and girls in different directions under the UK's education system (and probably many other countries too). Putting them in opposition I believe is unhelpful to say the least. This tension was manifest in the UK in the 19th century arguments between Matthew Arnold and Thomas Huxley, as well as the more recent and famously vitriolic exchanges between C.P. Snow and F.R. Leavis. And it continues. In the UK, where increasing emphasis is put on the STEM subjects at school under recent government drives, complaints not infrequently surface in the media. This can go beyond arguments regarding 'the two cultures', extending to the perceived maleness of the sciences. For instance, the journalist Cristina Odone complained that her daughter's school had a '*focus on STEM subjects* [which] *sends a message that makes her (and me) uncomfortable: doing a man's work is more impressive than doing a woman's.*'[38]

This journalist apparently believes that Arts are necessarily the appropriate subjects for women and STEM for men. I feel deeply uncomfortable with having stereotypes like that being promoted in the mainstream media, dividing subjects up into male and female and giving out the message that science is not for girls.

Policy makers can also fall into the trap of associating creativity only with the arts. In a debate in the House of Lords on careers' advice in 2018, Baroness Garden said '*there is a growing need for STEM skills, but not at the expense of creative skills*'.[39] Certainly, in the UK at least, these misplaced sentiments are widespread. Scientists must vocally reclaim creativity as part of their toolkit; a most important part at that. That is not to detract from the Arts, simply to recognize that creativity belongs to the scientist as much as the artist, and that both are appropriate for individuals of any gender.

It is interesting how words change in use, if not formally in definition. Design is another word that has become divorced from its engineering past to be part of the modern creative industries. This is a topic, along with creativity, that I have discussed with Christopher Frayling, former Chairman of the UK's Arts Council and Rector of the Royal College of Art. Given that background, it might be reasonable to presume he would come down firmly on the side of 'the Arts are creative and design belongs to them' school of thinking. Far from it. As long ago as 1993 he wrote:

> *Moving on to the designer, up until relatively recently the popular stereotype was rather different. Instead of the expressive artist, we have the pipe-smoking boffin who rolls up his sleeves (always his, incidentally) and gets down to some good, honest hands-on experimentation. From Leslie Howard in* The First of the Few *(1942), to Michael Redgrave in* The Dam Busters *(1955). The designer-boffin's very best moment of donnish understatement came in* The Dam Busters, *when the man from the ministry says to Dr. Barnes Wallis (Michael Redgrave): 'Do you really think the authorities would lend you a Wellington bomber, for tests? What possible argument could I put forward to get you a Wellington?' To which the boffin replies 'Well if you told them I designed it, d'you think that might help?' Cut to Barnes Wallis in the cockpit of a Wellington ...*[40]

Design as understood in its broadest sense is something appreciated by the tech firms in Silicon Valley. They recognize its fundamental importance in creating innovative new products, and they use the term (also applied more widely) 'design thinking' to encompass the core of the development of new products. As one of the key champions of this approach, Tim Brown, CEO of design firm IDEO, puts it in his book *Change by Design*,

> *Design thinking taps into capacities we all have but that are overlooked by more conventional problem-solving practices. It is not only human-centered; it is deeply human in and of itself. Design thinking relies on our ability to be intuitive, to recognize patterns, to construct ideas that have emotional meaning as well as functionality, to express ourselves in media other than words*

> *or symbols. Nobody wants to run a business based on feeling, intuition, and inspiration, but an overreliance on the rational and the analytical can be just as dangerous. The integrated approach at the core of the design process suggests a 'third way'.*[41]

Companies like Google, Amazon, and other tech giants know it is not sufficient to have scientists (or even 'boffins') working away in isolation without the social scientists, for instance, indicating what the buying public will relate well to and what is therefore likely to sell as opposed to bomb. Inventing something that misses the mark won't progress a company's bottom line. In this area it is clear there is no attempt to pit science and technology against arts because that is not the path to success.

Nevertheless, looking at the literature, it is depressing to realize how many late 20th and 21st century writers find it necessary to run down science instead of embracing it, to continue with the Leavis-Snow two cultures polarization in unhelpful ways. For instance, the American novelist Lucy Ellmann seems to think that in attempting to answer fundamental questions creativity is necessarily absent. She was quoted as saying in 2010: '*The purpose of artists is to ask the right questions, even if we don't find the answers, whereas the aim of science is to prove some dumb point.*' and, in the same article, that '*scientists are winning*'.[42] 'Dumb points' as she puts it, may just be what we need to solve some urgent societal challenge, and the route to achieving this is every bit as creative as putting paint on canvas or words on a page.

Of course, not all writers are narrow-minded about science. The Norwegian-American writer Siri Hustvedt has taken pains to study neuroscience and psychology to complement her writing. She believes that we need to look to science to explain creativity and that memory and imagination are two faces of a single process, with '*creative, original ideas rooted in memory—maybe not specific singular memories, but from the sum total of our experiences*'. She goes on to say:

> *It does not matter if one is doing physics or writing poetry. My favorite quote from Einstein addresses this question. Jacques Hadamard, the mathematician, asked Einstein how he worked, and he answered that his essential creative work had nothing to do with signs, either mathematical or linguistic. That part came later. His deep work, he maintained, was visual, muscular, and emotional.*[43]

By this point I'm sure I've made my point clear. Creativity belongs to all of us if we choose, whatever subjects we opt for at school and beyond. Science research abounds with it, and the next chapter will look at the driving force of curiosity. Either route, the Arts subjects or STEM, offer scope for those who wish to express their creativity, albeit in very different forms. However, there is no doubt that school education does impact directly on how children perceive where these subjects may take them and, as we have seen in the previous chapter, school, society, and parental influences all play an enormous part in the choices that are made and how boys and girls are steered in different ways, whether consciously or not. Telling young children that science is not creative is counterproductive and may lose STEM subjects many girls who have imbibed the message that girls *should* be creative.

Science Is for Everyone

It is my firm belief that science should be for everyone. Just as scientists should know their Shakespeare and be able to write good prose, so those who pursue careers in the arts or humanities should not be ignorant of some basic science to help inform their lives. Familiarity across the spectrum is healthy. This is not just a case of creativity, it is a case of being able to make sense of the world in a rounded way and so make good decisions at both the personal and societal level. We should not have to make a crude choice about which side of some perceived intellectual gulf we choose to sit on.

Historically this was not so difficult. Until around the mid-19th century, disciplines barely existed, and certainly were not drawn with rigid lines. Subsequently, not only did creativity start to be seen as the domain of the non-scientist but, as professions became more codified and 'scientist' became such a category, it became much harder to retain breadth. If we go back to Erasmus Darwin, he was able (and keen) to write a poem *The Loves of the Plants*, 1746 lines long, covering (in its original publication in 1789) 184 quarto pages discussing Linnean classification in which he described the 'sex lives of plants' in rhyming couplets. His near-contemporary, the painter Joseph Wright, notably depicted scientific scenes in paintings such as *An Experiment on a Bird in an Air Pump* and *A Philosopher Lecturing on the Orrery*. Another contemporary in Germany was Johann Wolfgang von Goethe who, best known as a writer, also spent much time on scientific pursuits; in the USA, John Audobon combined his interests in ornithology with exquisite and detailed paintings of the birds he studied. More recent examples of those who successfully merge science and the arts include the Austrian-born war-time beauty, Hollywood film star, and inventor Hedy Lamarr and her collaborator and co-inventor the American composer George Antheil. These two, perhaps unlikely, technologists were responsible for patenting methods which underpin modern Wi-Fi through frequency 'hopping'.[44]

Art and painting are, of course, associated with aesthetic beauty, but it should be remembered that science and, perhaps even more, mathematics, has its own form of beauty associated with it. Italian neurobiologist and Nobel Prize winner Rita Levi-Montalcini spelled this out explicitly when she said: *'I believe that my approach to science was from the point of view of the beauty of the nervous system and not just plainly because I was interested'.*[45]

This statement reinforces the importance of not trying to partition off certain attributes as pertaining only to either the arts or science. However, being a polymath with varied interests has become increasingly difficult, due to the need to specialize within the education systems of different countries (albeit the age of specialism varies somewhat), coupled with the idea that you have to make a 'choice' to sit on one side of the great divide or the other. Within the English system, if you try to 'mix and match' arts and science subjects at A level, you may find yourself ending up unqualified to be admitted to the more specialist science courses at university, even though it might have looked as if options were being kept open, and you were advised to that effect by teachers. For instance, to read Chemistry at an English university, typically both maths and a second science are required. In other countries, this early specialization is much less marked, meaning it is easier to retain broad subject choice up to 18 and even beyond. In the USA, for instance, a major does not have to be declared until well into a degree course.

Even for those who may have no interest in crossing domains or specializing in science and mathematics, nevertheless, some scientific understanding matters to help navigate everyday life. As the 2014 report *Vision for Science and Mathematics Education* from the

Royal Society (a report with which I was closely involved) on the future of scientific education expressed it:

> *Scientific discovery and technological innovation can provide solutions to challenges such as scarcity of food and water, energy supply and security and climate change, but they also raise social and ethical dilemmas. All citizens need the skills and knowledge to be able to make informed decisions about how society handles these issues.*[46]

In other words, this requisite knowledge is an issue fundamental to citizenship and democracy.

Here, I am not just talking about facts. Unlike C.P. Snow, who famously said in that much-quoted Rede lecture in the Cambridge University Senate House in 1959 describing cocktail party conversations,

> *Once or twice I have been provoked and have asked the company how many of them could describe the Second Law of Thermodynamics. The response was cold: it was also negative. Yet I was asking something which is the scientific equivalent of: Have you read a work of Shakespeare's?*[47]

What matters is not detailed factual knowledge but a general understanding of the fundamental principles, and being aware that there are such things as physical laws which simply cannot be contravened. I would much prefer everyone to remember the second law of thermodynamics via the words of the song by Flanders and Swann: '*Heat won't pass from a cooler to a hotter. You can try it if you like but you far better notter!*' rather than be able to produce a textbook version of the law. What matters is that everyone should know enough to interpret the facts presented and understand their relevance when making decisions about anything from the safety of vaccines to what decarbonizing our economy might mean for them in the interests of preventing global warming. They also need to know when 'facts' are actually being misrepresented, perhaps for political advantage, so the ability to read material critically is also a key skill.

The focus on textbook answers to questions for exam preparation is often what stifles a child's interest in science in school. It can, regrettably, convey the message that science isn't creative because there is always a 'right' answer; this, I suspect, is what Ellmann means by 'proving some dumb point' because she has no understanding of how science is done in practice. Rarely is it possible in a school to allow a child simply to explore some natural phenomenon. This is a great loss since children are inherently curious. Too often that curiosity fades as they get used to the formality of exam-driven lessons or are frustrated by their parent's inability or reluctance to answer their endless questions. As Nancy Rothwell put it to me when I interviewed her:

> *I worry that we increasingly teach science out of textbooks. Someone once said that's like learning to drive a car from reading of a book. It's the practical science that really turned me on, doing it myself. That was it, I was committed.*[48]

It is a shame our education systems too often prevent the natural curiosity developing into a lifetime's interest in scientific matters. Curiosity, as we will see in the next chapter, is a key part of what drives a scientist.

CHAPTER 6

BECOMING A SCIENTIST

It's important to understand what drives you .... We feel passionate about what we do. We love science, research and development. We are good at what we do. We work with the right people. We have our micro and macro environments. And we have goals that we believe we can achieve and that are worth getting up and working for every day.

Turkish-born German oncologist Ugur Sahin,
BioNTech Co-Founder with his Turkish-born immunologist
and physician wife Özlem Türeci[1]

I have four tips:

1. Curiosity. Go after your curiosity.
2. More curiosity.
3. Even more curiosity.
4. Passion. It's not enough to be curious—one has to really love what one does. For men and women, science is demanding and there are many, many dark periods, low periods.

Ada Yonath, Israeli crystallographer and Nobel Prize winner
in Chemistry, 2009[2]

She was the one who told me that I could make it even though I was a woman, and she did warn me that the road ahead for women in science might be more difficult, but not to be deterred

US physicist Millie Dresselhaus on advice given to her by
Nobel Prize winner Rosalind Yalow[3]

Progression

We have seen how the early years' environment may impact on girls. As education continues, additional effects come into play, which may reinforce the message that the STEM subjects, particularly the physical sciences, are not for them. There is a well-established field of study in social psychological research which studies the generic situation of being 'other', being in a minority, and the consequences for minority–majority interactions.[4] Although this body of work is concerned with a range of different contexts, it describes many of the situations women in STEM face and even more so women of colour. It highlights that, simply by virtue of being in a minority, the interface with one's working environment may feel very different for women from that for men, even with no ill intent from those around.

At every level, from schooling through to research, an individual who doubts their self-efficacy[5] and has a real fear of failure is likely to struggle compared with their less troubled peers. A low belief in self-efficacy when it comes to maths, a feeling closely related to 'math anxiety', feeds into a belief that mathematical subjects, particularly the physical sciences and engineering, may not be a wise choice.[6] This low self-efficacy, stemming from stereotypes and cultural norms which a growing girl imbibes from the world around them, is detrimental to encouraging girls to enter science. As one paper put it: *'we need to counteract gender stereotypical competence beliefs and assure women that they have what it takes to handle STEM careers'.*[7]

These studies point to the fact that there are many reasons, both arising from their childhood and adolescence and from the environment in which students and early career researchers find themselves, which build up to contribute to the observed decrease in the percentage of women at increasing degrees of seniority. Figure 3 shows an example taken for the early career stages averaged across the world, plotted by the UNESCO Institute of Statistics.[8] This figure demonstrates clearly how globally more women than men enter higher education in the STEM disciplines, but the fall-off is very rapid. Taking data from a single country, but extending much further across the academic hierarchy, Figure 4 shows data for US universities in biological and life science departments, showing a similar shaped graph, colloquially known as a scissor graph. The precise details of these scissor graphs vary by country and by discipline, but their shape

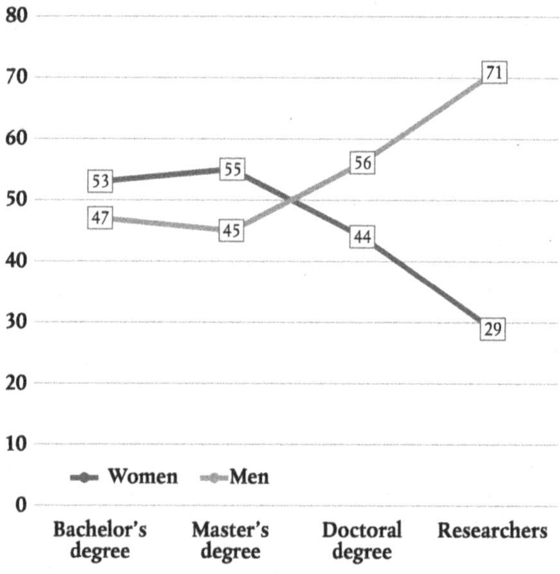

Figure 3 Global proportion of women and men in science as graduates and researchers, 2018.

UNESCO data for Sustainable Development, Institute for Statistics Blog 'Data + Policy = Action on International Day for Women and Girls in Science', 14 May 2021, http://uis.unesco.org/en/blog/data-policy-action-international-day-women-and-girls-science. Used with permission.

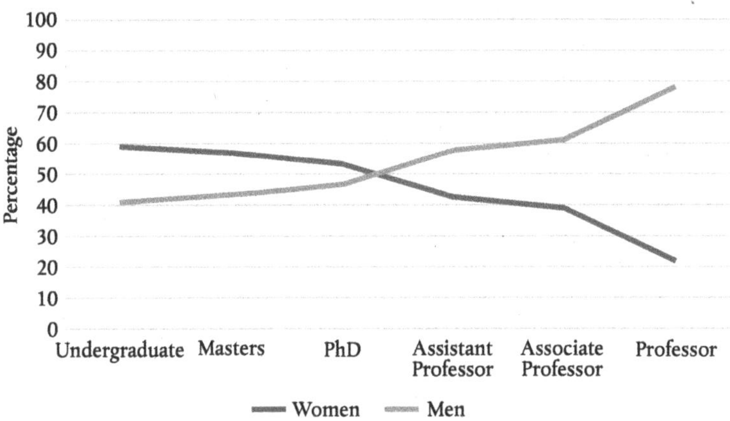

Figure 4 Percentage of women at different career stages for US universities for biological and life sciences, student data from 2014 and academic data from 2015.
'Gender equity: Addressing recruitment at the departmental level', *Inside eLife*, 1 November 2018.
Data from (USA) National Center for Science and Engineering Statistics, 2017. Used with permission.

will be, broadly speaking, similar, albeit in some subjects (for instance computing or engineering) the percentages of women start off so low at undergraduate level, the lines never cross, merely increasingly diverge.

Overall, the challenges that *any* aspiring scientist is going to face may feel significantly bigger for a woman (or a person of colour, or indeed for anyone in a minority in the workplace). Any book which details what it takes to be a successful scientist, if written by a (white) man, is likely to underplay the challenges as women perceive them. Tales from successful men illustrate that for them there may well have been hurdles that nearly derailed them. Too many women will find that derailment complete, so that they never achieve their full potential. Everyone is the loser in this case.

The Ingredients Needed

Curiosity, the desire to 'play', to take things in new directions, to be willing to take risks, to be speculative and imaginative; as the last chapter highlighted, these are all crucial ingredients in the make-up of a successful research scientist. Curiosity certainly sits at the heart of the necessary skillset, but it is far from sufficient. As the quotation at the head of this chapter from Ada Yonath spells out, any researcher will have long periods when things do not go right and nothing falls into place. For some, that can be overwhelming, or circumstances then mean future job security is lost. Resilience is a key requirement, but however tough one may be mentally, if the funding dries up it may be impossible to keep going. But that's assuming the novice is on the path (and wants to be) for a full-on research career. If things completely falter at this stage, the Nobel Prize may be out of reach but that doesn't mean that a successful career utilizing science isn't out there for the taking, as I've argued in earlier chapters. Failure in cutting-edge research—in terms of producing breakthrough results or progression up the academic ladder—does not mean failure in life, however much PhD supervisors may imply otherwise. Nevertheless, my focus in this chapter is on the ingredients needed to succeed in research (although not necessarily academic research), even if any given individual peels off that career trajectory sooner or later.

For each of these attributes, it is important to recognize that how they are perceived and need to be executed may differ for women from men, due to societal pressures and expectations, in ways that are subtle and, sometimes, not so subtle. The minority status of women in many labs, coupled with the possibility, or even probability, that some of their colleagues may have views of them locked into some inappropriate and/or outdated stereotype may mean that progress through a career can feel harder for women than for men.[9] Additionally, women have been shown to be more likely to be on the receiving end of what one study described as 'ostracism and incivility', largely (although not entirely) coming from men, and this inevitably has negative outcomes for women, affecting their psychological well-being.[10] But the issues around self-efficacy mean that it is also the case that stereotypes may colour women's own perceptions of themselves, particularly if an unsupportive environment piles on top of their anxieties, thereby causing them to be diffident or down-beat in ways that hold them back. Still, practice makes perfect: resilience can be built over time, but it does need more determination to keep going, something many men will not experience in the same way. As one former head of a physical science department and dean said to me privately, when looking back on her earlier years as a researcher '*It was tough to be seen as different, and treated as such too, whereas with practice and experience I just tended to get used to ignoring or dealing with that*'.

For all scientists, many factors come into play when trying to find a path through the research maze. Luck is certainly one of these, however uncontrollable: luck in finding just the right research topic; luck in finding inspiring and supportive collaborators or mentors; luck in being in the right place at the right time, and so on. Confidence also matters. Not simply confidence in being prepared to stand up and deliver a talk in public—though that helps—but confidence to keep going with testing hypotheses and ideas and, on occasion, confidence not to let doubters derail unconventional ideas. Who you are surrounded by is likely to impact on how easily that confidence builds or shrinks. The evidence shows that, for women in STEM studying for PhDs, if there are few women around them, the probability of them failing to complete their studies starkly increases.[11]

Science, like scientists, comes in many forms. This diversity, as we have seen, is a huge benefit for science overall. For an individual to find exactly the right area about which they are passionate, and which is ideally suited to their inherent skills, may seem a bit hit and miss, but without being fired up, success is less likely. Frequently, but certainly not inevitably, young children and teenagers work out early on what fires them, so that they can be single-minded about it. Others take more time to find their *forte*. Frances Arnold, as we saw in Chapter 3, changed her discipline quite radically, from aeroengineering to biochemistry.

Certainly not everyone works out straight away what suits them. Sometimes, even after they are well underway with research, they may make a decision to switch fields because they feel they need to, in order to get a better grip on the first question under study—or because they feel their own field is stagnating. This was how Nancy Rothwell, President and Vice Chancellor of Manchester University, explained it to me when I interviewed her a few years back.

> *I felt the field* [of obesity] *was stagnating. I also believed that the real breakthrough in the field was going to be when the gene that causes genetic obesity was found, and I'm not a geneticist, I'm not a molecular biologist ... I thought I am not going to make the breakthrough in that field. I need to go into another.*[12]

So she did. Having already spent around a decade in that first field, she went on to have an extremely successful career studying the role of inflammation in brain disease. As I mentioned earlier, I too changed fields, although for less high-minded reasons, but it was the making of my research career, and was certainly down to luck not judgement. Finding the right niche is crucial. If you don't have support around you, often the case for women, having the confidence to make that move is bound to be much, much harder.

The 'ingredients' I am referring to here, reflect not just the intrinsic skill set a scientist needs, but how these interact with externalities, such as those people around them, which is where gender comes into play. Exploring the different aspects needed in the scientist's armoury will demonstrate some of the stumbling blocks that can knock an aspiring scientist back, perhaps to the point of giving up. The interface with others highlights why and where, on average, the environment can be tougher for many women compared with their male counterparts.

Failure and Resilience

Let's turn to the dark periods of Yonath's quote. Not everyone would agree with the conclusion regarding setbacks from Iraqi-born clinical geneticist Lihadh Al-Gazali when she says: '*My advice to young women scientists is not* [to] *be put off by obstacles or a few failures. One or two setbacks are good for the soul*'.[13] Nevertheless, it is fundamentally important to be able to come back from setbacks if progress is to be made. They beset every scientist at some point in their career, and probably repeatedly so. Whether a Nobel Prize winner, or someone far more humble and much earlier in their career, finding ways to cope with the inevitable frustration is a vital attribute. When the experiments fail or give you unexpected results, that's when you need to be creative:

find better methodologies or rethink your hypotheses; think laterally. However, the disappointment and frustration that arise when things go wrong time after time can lead many to give up. No one's research ever goes right every time. Having a growth mindset is likely to help overcome the inevitable frustrations by encouraging the belief that hard work can make a difference.

I've already noted that children as young as seven recognize activities that you need to be 'really, really smart' to be able to do. In adult life, physics, economics, and philosophy are all disciplines that get labelled with the adult equivalent 'brilliance' label. How hard will you push yourself when things go wrong if you believe you have to be brilliant, but actually don't believe you are (correctly or not)? How will that colour your reactions, and what can be done? The fixed mindset perspective implies that abilities are innate and unchangeable, so there is no point in trying, whereas those who believe that growth is possible feel that with appropriate effort and practice, progress can be made. These ideas feed into early years education, but carry on once students are at university.

A study from the US published in 2011 explicitly tested the hypothesis that the percentage of women PhD students across both STEM disciplines and humanities and social sciences would be inversely correlated with how brilliant people (men and women) believe you have to be to succeed in these subjects. In other words, whether people believe that success is down to innate talent. Other hypotheses were tested too, including whether success in these disciplines is down to sheer hard work, but the only strong correlation found was with the concept of brilliance, as shown in Figure 5. Maths, Physics, and Computer Science all score highly on the 'brilliance' factor[14] and have very

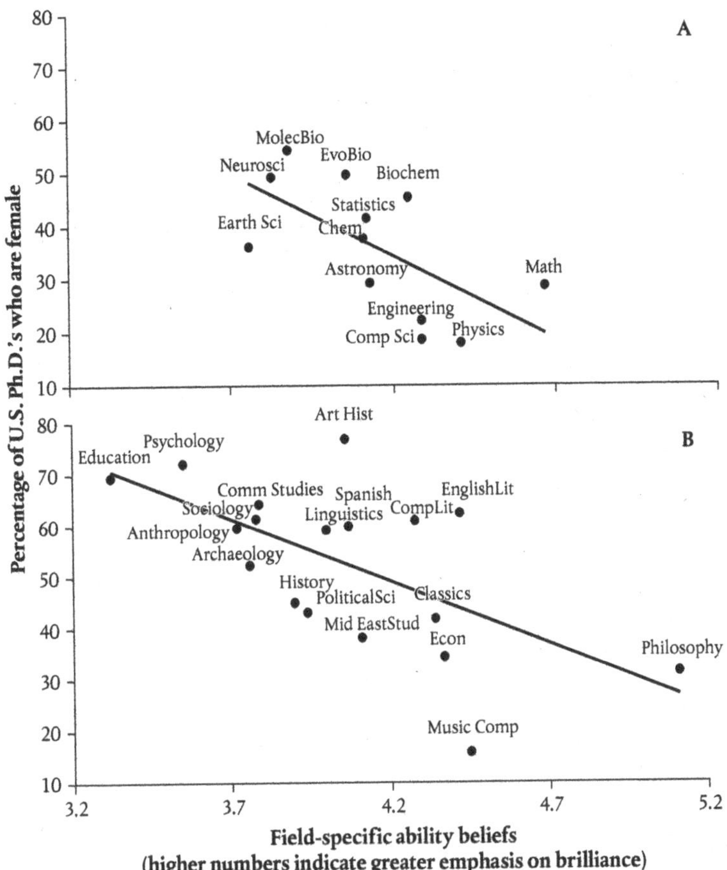

Figure 5 Field-specific ability beliefs and the percentage of female 2011 American PhDs in A) STEM and B) Social Science and Humanities.
Sarah-Jane Leslie, Andrei Cimpian, Meredith Meyer, and Edward Freeland, 'Expectations of brilliance underlie gender distributions across academic disciplines', *Science* 347 (2015): pp. 262–7, doi: 10.1126/science.1261375. Used with permission.

low numbers of women studying for PhDs (the same is true of Economics and Philosophy), whereas the Biological Sciences have much more nearly equal numbers of men and women, and these are subjects less strongly identified with innate brilliance.

The same authors found very similar effects when the percentages of African Americans in different fields of study were examined, but not when Asian Americans were considered.[15] Those beliefs that are already present in seven-year-olds, about boys being 'really, really smart', whereas girls are not, play out years later in the subjects in which adults choose to specialize in research.

Considering the impact of mindset in attitudes towards STEM studies, one study published in 2022 specifically looked at how men's and women's attitudes changed with regard to their own strengths during a semester-long introductory Physics course in the US.[16] The women were more likely than the men to believe that physics requires innate ability, and also that that was something they lacked; this negative feeling regarding their own competence grew during the course. The study highlighted how instructor interventions facilitating a growth mindset could offset this feeling. How we feel internally, induced by everything in our lives so far from parents to school, from reading books to watching films, will impact on how we cope with setbacks.

Failure is a fact of scientific life, although it may take many forms. It may be that an idea is just wrong. Alternatively, sometimes it is impossible to devise an experiment able to prove definitively that an idea is right, simply that it might be, a limbo that is intensely frustrating. Perhaps the experimental approach doesn't have the sensitivity to deliver the necessary results. Or, as Darwin feared when contacted by Alfred Russel Wallace, who was reaching similar conclusions to him regarding evolution, a researcher can be scooped, particularly when the problem is in some topical area. Perhaps the researcher's skills really aren't up to the problem, as I found to my cost when trying to get to grips with three-dimensional thinking about grain boundaries in metals in my first postdoc. In this case, I was working in an area which did not play to my strengths.

There is no doubt, however, that it is possible to learn from failure, as long as the individual can pick themselves up thereafter. As the Lebanese-American physicist Ani Aprahamian put it: '*After my 23 years as a tenured professor, I can see that nothing I did was a wasted effort. I learnt from failures as much if not more than my successes*'.[17] When I was asked to give a TEDx Whitehall talk at the Royal Society in 2018, I gave my talk the title 'Nothing is wasted—turning negative experiences into positive life lessons'.[18] Giving this talk was nerve-wracking enough, but circumstances made it even worse. Five minutes into the talk the sound engineer came up to tell me my microphone wasn't working properly and would I start again with a different unit. I refused to do this but continued with the new unit (the change in sound quality is very obvious from the video), saying that from an experience like this you might learn that '*you never want to do another TEDx talk*'. That got me a laugh but had that 'failure' happened twenty or thirty years earlier I'm sure I would not have had the aplomb and confidence to crack such a joke. It reinforced my overall message, unfortunate though the occurrence was. As in my case, a woman's confidence may grow with the realization, over time, that one failure need not destroy an entire career.

Persistence and resilience are necessary characteristics of the research scientist if they are going to keep going. Marie Skłodowska Curie demonstrated this in abundance, as has just about every successful scientist, male or female, since. But that is not to say it is easy. As she put it:

> *At times, I would be encouraged by a little unhoped-for success; at others, I would be in the deepest despair because of accidents and failures resulting from my inexperience.*[19]

Venki described the issue of courage to keep going in a 2013 interview:

> *It takes a certain amount of courage to tackle very hard problems in science, I now realise. You don't know what the timescale of your work will be: decades or only a few years. Or your approach may be fatally flawed and doomed to fail. Or you could get scooped just as you are finalising your work. It is very stressful.*[20]

Even for those who ultimately become Nobel Prize winners, like him, failure was always on the cards, but it is easier to look back with equanimity on previous failures if success has followed. Women, as outsiders, may find courage in short supply, particularly if they believe they are being more closely scrutinized than their male peers, provoking a tendency to question themselves. Those who have got successfully onto the academic ladder may still find themselves wondering how their own self-image fits with what others seem to expect and hence find themselves struggling to reconcile what they think is expected, with who they think they are. To quote the words written to me by a former student of mine, Beth Bromley, now a mid-career academic at Durham University: '*I don't feel like an established academic, I'm not sure I ever will. I don't fit my own internalized model of an established academic, which may well be unattainable.*'

Impostor Syndrome

Impostor syndrome is another form of lack of confidence that is to be found in many a scientist, however externally confident, and especially found in women; to some extent that last quote explains exactly how feeling an outsider feeds into impostor syndrome. Defined as a feeling of being a fraud who is in danger of being found out, impostor syndrome is certainly not restricted to science and scientists. It leads to a sense of not genuinely belonging and a worry that, for instance, speaking up is bound to lead to trouble. Jocelyn Bell Burnell described in an interview with me how she felt when she first arrived in Cambridge to start her PhD, neatly summing up how she was assailed by impostor syndrome:

> *Looking back I realise I was suffering from impostor syndrome. I'd been in Northern Ireland, York, and Glasgow, all in the far north west, and as a real country bumpkin I'd come to this terribly suave place,* [people were] *frightfully confident. Some people were terribly keen to tell you how bright they were. And I thought,* [a growl], *I'm not that bright, they've made a mistake admitting me, they're going to discover their mistake, and they're going to throw me out.... but I'm a bit of fighter ... so until they throw me out, I'm going to work my hardest, so that when they throw me out I won't have a guilty conscience. I'll know I've done my best.*[21]

Impostor syndrome is a familiar shadow in many an apparently successful scientist's life, as Donna Strickland made clear when she was asked whether she had ever suffered from it '*It's hard to believe that anyone hasn't had to deal with that.*'[22]

Despite women apparently being more prone to be afflicted, or at least more willing to talk about it, it's important not to let that inner impostor win. That may be easier said than done. A key part of being able to overcome the syndrome is to see it for what it is. Coupled with the very present danger of disagreement that the requirement of presenting results in public exposes—a core part of a research scientist's activities, be it at a conference or simply a departmental meeting—it can lead to substantial 'under-selling' of their work by women. Add in the possibility of bias as well, to be discussed in the next chapter, and the net effect is that women may see their discoveries attributed to another (male) scientist, the Matilda Effect (in contrast with its converse, the Matthew Effect) discussed in Chapter 2. Of course, a woman who fails to receive due credit for her discoveries is unlikely to make the progress she deserves. As Sally Davies, the UK's former Chief Medical Officer and current Master of Trinity College in Cambridge expressed it:

> *Well, like most women of my age I've had to be better than the men to get the jobs, and then of course there is the inevitable: you make and put forward an idea at a meeting and it's only when a man says it that it's taken up. So, I've faced all the challenges women have along with my own touch of the impostor syndrome – we all have it.*[23]

Like Donna Strickland, this is an extremely powerful woman admitting to impostor syndrome, and also demonstrating the feeling that women often get short shrift in comparison with their male colleagues.

I first learned about impostor syndrome when participating in a public conversation with Alison Richard, the then Vice Chancellor of Cambridge University, in front of an audience of early career women from industry and the University. So it was rather late in my professional career before the term itself entered my lexicon, although the emotion had long been lurking. It is good that these days it is talked about so openly, so that people know they are not alone in their impostor feelings. Alison told me how she had assumed, when she got her acceptance letter from Newnham College to be admitted as an undergraduate, that it was a 'clerical error', as she put it. A similar emotion was expressed in a talk I heard around the same time by science writer and journalist

Vivienne Parry, who expressed her astonishment when she first received a key to what was usually a private space in the BBC's building, feeling there must have been some mistake, but she'd better use the key before it was taken away from her.

It was helpful to put a name to a feeling that I had often felt; I had not appreciated how common it was. I have learned, any time I have to take on some new role and feel a fraud lacking any track record in that space, to use the emotion in the same way as actors so often use stage fright: as a source of adrenaline to get through the experience, recognizing that if I don't feel that anxiety I may well not give my best 'performance' in whatever new role it is. And, having survived that first occasion, to build on it thereafter, as I spelled out in my TEDx Whitehall talk.

Sometimes, however, it is the external circumstances that make it clear other people think you're an impostor. I can still recall the first grant-giving committee on which I was called upon to serve, which would have been at least twenty years before I came across the explicit concept itself. I was uncertain what to expect, uncertain about my standing too, not least because I had just left my baby son for the day at a time when I was still breastfeeding, so I was nervous on many fronts. When I arrived at the committee room I was immediately accosted by some grey-suited elderly gentleman complaining how late the committee papers (and in those days, they were papers, huge piles of them) had gone out. I looked at him rather blankly before realizing he had simply assumed that, as a woman, I was part of the administrative support, not a member of the committee. It was not a good beginning, although I'd like to think he soon learned how wrong he was, once we started discussing the grants. I fear most women scientists could give their own examples, probably multiple examples, of similar assumptions about our role.

Other People: Mentors and Sponsors

There are many factors beyond passion and a willingness to take risks that contribute to the ability to keep going, and crucial amongst these are the roles other people play in a researcher's life, as well as that fickle entity: luck. Many people, in different guises, matter in a researcher's life. I've read several memoirs and autobiographical accounts by scientists who manage to omit their research students (and postdocs) more or less completely and, to my mind, inappropriately. Beyond a certain point in one's career, it is all but impossible to keep going with the hands-on science. My research has, as I've progressed up the ladder to take on other responsibilities such as teaching, administration, and committee work, necessarily been done by others for the past 25–30 years, and specifically those whose research I've supervised. While I have tried to advise, encourage, stimulate, feed in ideas and tell them when things are heading off track, it is a very different skill from doing it oneself. In my case this shift away from research certainly led to a form of withdrawal symptoms at the time: I missed doing the hands-on experiments, the excitement of staring down a microscope, waiting for the veil to be lifted. After a while, though, I knew I was no longer up to date with the techniques and apparatus, and there was little point in longing to get back to the bench. It could only end in (metaphorical or literal) disaster.

Students and postdocs become, of necessity, the medium through which ideas can be explored. Many group leaders in the later stages of their careers remark that watching individuals in their teams develop into outstanding researchers themselves is one of the most rewarding things they do, once hands-on research is effectively denied to them. The supervisor must impart their own wisdom and experimental or theoretical/computational green fingers to the novices in their care. It is a sad fact that people get promoted in large part because of their own first-person skills in research, and then they have to stop working that way. Instead, they are supposed to acquire a completely new set of skills, often with little training or even advice, in order to achieve these aims. That is a stage when mentoring becomes so important, a skill which the university sector is certainly not good at providing formally. That some professors not only intrinsically lack these skills—and no one automatically points this out to them—but are also bad at communication or prone to bully, is another systemic problem.

In reverse, it can be seen that for the researcher coming up through the system, the role of their supervisor is crucial. If this individual sees a female researcher as less worthy of their attention than their male counterparts, or actively denigrates them, then that person's confidence will be sapped to the point they may leave the field completely. In some cases, the behaviour of a supervisor towards female members of their team can become totally unacceptable; harassment is not uncommon and is one of the many gendered slings and arrows I will address in the next chapter. I have benefitted from some wonderful mentors, but I have also worked with others who did not give me much support.

Mentors are those who provide professional support and encouragement and give advice. They may be a line manager, such as a PhD supervisor, and sometimes they are formally provided through a departmental scheme. When the relationship works, a fantastic and long-lasting relationship can form. I'll get very personal when talking about the mentors who made me what I am, because it is at the personal level that mentors so often influence an individual. However, an assigned mentor through a department or other scheme may turn out to be a dead loss; there is an element of chance in the pairing, and finding a mentor for yourself, by whatever route, may be at least as likely to be beneficial as any formally assigned.

My first key mentor was my (second) postdoctoral supervisor, an American called Ed Kramer. He remained a mentor right up to his death a few years ago. It wasn't enough for Ed that I was flourishing scientifically; he wanted to make sure I got known by others in the field. To this end he made sure I got invitations to give seminars in various universities around the USA. He introduced me to the top professors in the field when we went to conferences and made sure I joined them for dinner so that they would remember me. (He did the same for outstanding male postdocs as well.) Looking back, I wonder if other professors thought it odd that Ed brought along this young woman, when they were still very thin on the ground. This appeared to have bothered him not at all. I was simply someone with whom he could discuss and dissect theories, models and wild ideas at length.

He turned me from a no-hoper into someone who could go on and thrive in academia. He did this within weeks of me starting to work with him, after two years of getting nowhere in my first postdoc position, when I lost all my enthusiasm and love for science. He gave me confidence and contacts—not to mention that the project gave rise to an enormous number of publications. For the rest of his life, he was always there in the background giving me encouragement. He was not a great letter-writer (or its updated incarnation as email), so contact across the Atlantic was spasmodic in later years. Nevertheless, if I needed advice or a reference letter, Ed was always more than willing to step in. His importance in my life and its successes were, in large part, down to luck: he needed a postdoc at the time I was in the department anxiously trying to find a position to enable me to stay in the USA. Unfortunately, it is not that common to stumble upon such a wonderful relationship and far too many people never are able to forge anything similar.

Mentors don't have to be in any sort of line-manager role, they may simply be colleagues encountered in a department or at a conference. Although many women seem to think that they can *only* thrive with a female mentor who can relate to their specific challenges, I don't subscribe to that view. As it happens my key mentors have been men, not least because women have been few and far between in my field. There may certainly be occasions when the female view may seem to be more relevant (although my experience described below shows even that is not necessarily true), but anyone who takes a close interest in an individual and is willing to expend the time to help can serve in a mentoring capacity. They may be relationships of many years' duration, as Ed's was for me, or relatively short in span to cover some critical juncture in life. I have mentored people in both ways myself. Undoubtedly, there is satisfaction in the act of mentoring, as the computational biologist Sarah Teichmann describes:

> *I wouldn't be able to look at myself in the mirror if I was trampling on anyone. I have had fantastic experiences with my mentors and I want to be that person for my group members, irrespective of whether it makes me more successful or not.*[24]

There is no doubt that mentoring takes a particular kind of interpersonal relationship to develop to work well, but also there is a requirement that the more senior person can identify with, or at least recognize, the issue(s) that the junior is working through and wanting advice on. For this reason, successful mentoring is probably particularly difficult in cases where intersectionality comes into play. A Black woman, or a woman with a specific disability, may find it exceptionally hard to find someone who can assist them with the problems they face, although there may be plenty of people who would be willing to try to help. But, given that white women without a disability find it hard enough to find a useful mentor, it is inevitable that the situation is significantly worse for those for whom other factors are also present. For senior management simply to nominate someone to act as mentor, and assume all will be well, is almost always going to be naïve and can amount to no more than a box-ticking exercise of little value.

In many circumstances, peer-to-peer mentoring also has much to offer; advice from someone perhaps just a year or two older who has been through the same challenges and whose memory of something to be faced down is fresh. It could be something as relatively ordinary as a first conference talk, which a professor may well take in their stride so comfortably it is hard for them to recall the horrors of the first occasion. A peer who did it only a year or two previously will probably be able to enter into the fears of the novice, and hence offer constructive advice, much more readily.

Mentors are important in any position and at any stage in a career, the more so if you are feeling isolated as a minority researcher not quite fitting in, as is so often the case. Anne Glover, a Scottish biologist who went on to become the Chief Scientific Advisor to the European Commission, spells this out:

> *The most important mentors in my life have been men, because they have explained to me how the system works. If you are not part of the club, you*

> *don't know how decisions are made, you don't know how people get on the important committees, and so on.*[25]

There is, in science just as much as anywhere else, a lot more to success than the merely technical. Being able to be persuasive, confident to speak up but not so arrogant that people turn their backs, able to contribute to 'important committees' and knowing how the system works, are all important skills that need to be picked up along the way. Mentors are vital in this space; often more than one may be needed, each with their own strengths and weaknesses.

Along with Ed Kramer, the second key mentor in my life was the Welsh physicist Sam Edwards. You might think discussing the timing of a pregnancy was not something to discuss with a man, let alone one's Head of Department, but a crucial conversation I had with this man proves that such assumptions should not be made too easily. I will never forget the conversation I had with him at the time I was hoping to apply for a recently vacated Lectureship, while I was still on a fixed term position as a research Fellow. This was crunch time for me in my personal life. My husband and I knew we wanted to start a family (I was already over 30) and I let Sam know this. If he, as head of department at the time, had turned away and said effectively that if I wanted a family I shouldn't expect to be taken seriously as an academic, I have no idea what I would have done. This would have been in 1984 or 1985. Such a response would hardly have been surprising at that date (regrettably, that can still be the case in some quarters and the 'mother penalty' is still very real as we will see). But, on the contrary, what he said was '*intelligent women should have families*' whilst simultaneously encouraging me to apply for the Lectureship. I did apply, and successfully so. I was pregnant by the time I took up my lectureship at the start of the academic year 1985–6, although not even I knew that at the time. Becoming the first woman lecturer in the department, and giving birth in that first year, might have led to many negative comments but, if such occurred, they did not make their way to my ears. Another instance of luck in my life.

Sponsors differ from mentors, although often that distinction is not clearly made. The level of interaction may be much less, but crucially they give their protégé exposure to senior academics who may be of benefit, and they make sure that these people are considered for promotions, prizes and so on. This is how what is colloquially known as the 'old boy's club' operates. In hindsight it is clear to me that both Ed Kramer, and even more so Sam Edwards, played this role in my life as well as mentoring me, encouraging me and putting me forward for awards and promotion. These sorts of interactions are intangible, perhaps often invisible, but none the less important for that, and traditionally have worked more easily for men, as highlighted by one recent study looking at the international situation in the STEM disciplines.[26] The results showed that men were more likely than women to receive sponsorship from senior men; most women indicated that access to senior academics who could make a difference was not easy. The authors concluded that this difference in sponsorship perpetuates male dominance in STEM, conferring an invisible advantage on men.

If bright early career researchers do not know about job or fellowship opportunities, they are unlikely to put themselves forward. And, at a later stage, knowing what is going on is equally important so that you can apply for key positions that will enable your credentials to be recognized. Once again, Anne Glover recognizes one of the social handicaps women may operate under:

> *In the School of Informatics in Edinburgh, when they built a new building, they decided to have unisex toilets, because when guys go to the toilet they often chat with each other about departmental politics. A conversation might start as follows: 'I am coming off the finance committee, are you interested in that? Can I put your name forward?' As women are not in the toilet, their names are not put forward. So, either you have unisex toilets—and some people are not so happy with that idea—or you ensure that there is really good governance and transparent procedures to select candidates.*[27]

I have likewise heard of similar conversations along the lines of '*would you be interested in applying to be [University's X] next Vice Chancellor?*' taking place in a urinal. These may not be deliberate exclusions, but they are exclusions nevertheless. Opportunities for sponsorship are lost; women do not have this access.

Other People: Collaborators and Teamwork

As I have said earlier, few areas now progress via the work of a single scientist; teamwork is increasingly the name of the game. But it is so much easier to work well in a team when there is pleasure in the interactions. As Venki puts it:

> *Collaborations work best when the people involved are good friends, enjoy working together, and have complete confidence in each other, or when they have complementary expertise that allows them to tackle something neither could do alone. They also require a willingness to give up complete control of the project and to share credit in ways that might not seem fair to all parties.*[28]

The converse of this can be seen in Rosalind Franklin's experiences with Crick, Watson, and Wilkins, where there was mutual distrust and hostility. Some of this mutual antipathy seems to have had little to do with science and more to do with character—and misogyny. Simon Altmann, an Argentinian PhD student in theoretical physics at Kings College, London, at the time, describes the gulf between her and her colleagues, with Franklin being:

> *very well read in two languages, who was used to a civilised intellectual life, discussing painting, poetry, theatre, and existentialism. Now she found herself among people who had never heard of Sartre, whose chief reading was the Evening Standard, and who enjoyed 'the type of girls who would get drunk at departmental parties and be passed from lap to lap having their bra undone'.*[29]

This was not an atmosphere where trust was likely to develop or, referring back to Venki's words, one in which Franklin was likely to enjoy working with Wilkins and those around him. It is hard to imagine most women would have found that atmosphere congenial, then or now. There was also the issue of scientific 'ownership'. The head of the laboratory, John Randall, had indicated to Franklin that she would be in charge of the DNA work, but did not pass this on to Wilkins, who thought it was 'his'. In any laboratory, such a lack of clarity might be expected to lead to trouble. Certainly, such misunderstandings are not going to build trust.

I cite the case of what went wrong for Franklin because it is so well documented and exemplifies how *not* to have a successful collaboration. If you suspect a collaborator might steal results (as in essence Crick and Watson did), or looks down on you (as seems mutually to have been the case between Wilkins and

Franklin as well as with Watson), who you are clear can't see you as a scientist so much as an irritating and angry (but not pretty) woman (again illustrated by Watson's comments) or you don't want to spend time thrashing through ideas, scientific and more, with the person, then it is not likely to be productive. As Nusslein Vorhard put it: '*I once had a professor who thought women were better suited for pottery classes than for a lab. His remarks were extremely insulting*'.[30]

Attitudes demeaning women are not uncommon from male colleagues. Gina Rippon, a retired British neuroscientist, told me that as more women were appointed in her department they were collectively referred to as 'the girls'. I myself have been accused of having a 'hen party' when talking with a group of women in my department.

I consider myself fortunate that I've had some wonderful collaborations with people I do respect and admire, people I trust and people I enjoy chatting to at the end of a long day of analysing results. Sometimes these collaborations grow from supervisor/student or postdoc relationships once the latter has 'graduated'. I've had a number of these, sometimes picking up threads many years later. The requisite trust, one hopes, was built during those early years. My friend Richard Jones, whom I mentioned in an earlier chapter, I've known since he was a PhD student. In due course he became a lecturer in Cambridge and we worked together on a range of grants, co-authoring papers and thrashing through ideas between us even when they were extremely preliminary and possibly daft. The advantage of a trusted colleague is that one can bat wild ideas around with them without embarrassment. Additionally, we could discuss the whole way our joint group was run and how it fitted into the inevitable departmental politics.

Having trusted (scientific) friends, to discuss whatever preys on the mind, is as important in this sector as in any other professional sphere, but they have to be picked with care so that there is no threat of 'scooping' or of ridicule. Unfortunately for women, the issue of sexism may underlie interactions, or at least the worry may exist that this could be the case. A 2015 study looking specifically at the field of nanotechnology showed how the underrepresentation of women in STEM has an impact on the interpersonal processes of scientific collaboration, to the disadvantage of women scientists.[31] The results showed that for women, just over a third appeared to have no female collaborators at all. In contrast, for men a total of 80% of collaborations were with other men, and only two men reported no male collaborators.

There is relatively little work looking at patterns of collaboration around gender, although it is beginning to appear in the published literature. What is emerging is that there are likely to be disciplinary differences. Nanotechnology, as mentioned above for instance, will not be the same as genomics, another area where careful examination has been carried out and where it appears that there is a very substantial under-representation of women, despite the fact teams are large.[32] The authors of this latter study hypothesize that this arises because there are few women at the top of that field and, as it is the protégés of prominent scientists who are likely to be put forward for faculty positions, the under-representation of females in those labs propagates. In other words, as noted above, sponsors matter and are typically men sponsoring men. The authors also hypothesized that the environment in this specific field may be actively unwelcoming. More disciplinary analysis is clearly required to see if these arguments stack up in other fields.

Who you collaborate with matters. Clear evidence of 'homophily' amongst men (liking to work with those of the same gender; in this case, men collaborating solely with other men) was obtained in a 2021 study looking at data from Poland.[33] This effect was present across all STEM disciplines. Women were, however, much less likely to collaborate with other women, and this effect got stronger in research-intensive institutions. The lack of collaboration with men seen amongst women may arise if women believe men are the gatekeepers, so that being seen to collaborate with them is regarded as of more benefit than collaborating with women, who are presumed to be of lower status. However, there are standard social awkwardnesses to traverse too, as expressed privately to me by a senior Professor of Physics in the UK:

> *... when I attend, say, a large conference in the US, it is so much harder for me to approach a male group leader of my seniority level and try to invite them to go for dinner and chat about possible avenues for collaboration. They generally look quite concerned when approached, seem worried about being seen one-to-one with a woman and keep their distance during conversation. Yet I see males immediately switch to chummy business mode after they are introduced to one another.*

These social minefields cut both ways, as increasing numbers of women take on leadership roles, as she went on to say:

> *But I noticed at a recent* [funders' event], *when chatting with the Chair and a Deputy Chair, both female, how awkwardly a male participant tried to approach us, connect into the conversation, but then gave up and left. It seems that these problems are much the same in reverse, and perhaps this will*

prompt a re-think of single-gender advocacy as more women in STEM populate more senior roles, and males experience and understand these issues at first hand.

Communicating—How and with Whom?

Science, it should be clear by now, is not a lonely endeavour. It is not necessary to go as far as the French sociologist Bruno Latour in his book 'Laboratory Life'[34] and see the lab as a unit for sociological study by analysing the interpersonal activities as if with no prior knowledge of the structures, to recognize that human interactions play a large role. If the student–supervisor relationship—or that involving a more senior researcher, as with Franklin—breaks down, it can become extraordinarily difficult for all parties. If blame starts being parcelled out when equipment breaks down, the loss of trust in a team can be damaging for the whole project; a single awkward (in whatever sense) character or a blighted love affair in a group can sap morale more widely.

Beyond the actual doing of the science, it is necessary to stand up and let your voice be heard, as Glover's quote indicated above, be that within the lab team meeting, at a committee meeting, in a conference hall or in the media. Sometimes reaching out to the public is an important part of the science, so that results that matter to everyone can be disseminated. In other words, a final ingredient a good scientist needs is the ability to communicate, and that includes in any teaching they do. Be it in technical jargon with fellow scientists, or more colloquially with the public, school children, students, or politicians, if a scientist can't communicate effectively (and each of those audiences will need a different style) their work will simply gather dust. But beyond the imperative of communicating to share the work that has been done, or the more selfish desire to see one's own importance reflected by acceptance within the scientific hierarchy, is the importance that, particularly where public money is involved, the work is accessible to other researchers, not closeted away in some forgotten tome.

However, here too gender can come into play. The evidence regarding university teaching is clear: women are typically regarded as less competent.[35] This has been tested using student evaluations, removing all possibility of actual differences in performance by looking at scores for online teaching. In one study along these lines, students rated those presented as having male identity significantly higher than those presented as of female identity, regardless of the instructor's actual gender.[36] The words used to describe men and women also show significant differences in evaluations. A website has been set up to permit exploration of words used about male and female lecturers, as found in completed Rate my Professors[37] reports, compiled by discipline.[38] It's an amusing site to explore. To pull out some specific examples, it shows how women are more likely to be described as sweet and men as arrogant, regardless of discipline. Women engineers are more likely to be described as unpleasant than in any other discipline (and far more so than men), and female physicists and mathematicians are much more prone to be described as confused than their male counterparts. Even in a country perceived as one of the best for gender equality, Iceland, an analysis of student evaluations shows bias continues to exist: male students were found to rate female teachers lower than their male counterparts.[39] There was also a difference in the nature of comments each received. Whereas comments on male teachers referred to their subject knowledge, comments regarding women tended to be tied to their service to students and their relatability. Thus, stereotyping of roles persist even in such an apparently egalitarian society as Iceland.

Since student evaluations can feed into promotion criteria, this will be deleterious for women in practice, but it also points to the possibility that female scientists who try to be active public speakers may be disadvantaged. This hypothesis is borne out by the evidence. One study regarding perceptions of female communicators showed that they were likely to be stereotyped as 'bitchy', 'bossy', and 'emotional', often by other women.[40] Asking women to be role models by speaking publicly may have a high associated cost, on top of the negative feedback *any* public speaker may receive. Analysis has been carried out on scientists, men and women, who turn to YouTube to discuss scientific ideas.[41] Channels hosted by women were less popular on average. Worse, although the average female communicator received more comments per view, these contained larger proportions of comments that were hostile or sexual, and with more relating to appearance than for men. That this is so, is only likely to deter women from taking on these roles, thereby reinforcing the public impression that there are not many women in science. These findings regarding the disadvantages for women speaking publicly is yet another manifestation of what has been termed 'double trouble'.[42] In this, the combination of being in a minority in the workplace and working in a space where they are negatively stereotyped leads to high levels of gender-based threat. Gender-based threat, in turn, negatively predicts women's satisfaction and confidence in their careers.

This chapter may suggest that life is inevitably going to be difficult for every woman trying to pursue a career in science. I think that's far too negative a way to view it, and many (although manifestly, not all) women I've talked to say something along the lines of '*I have never felt excluded or disadvantaged in any way because of being female in any department I have worked in*', even if they then go on to discuss the frustrations they've felt in parts of their job. Women will always face challenges, as will men, but some of those may well be different from those that men experience. However, the problems described here are often inadvertent, tied into old-fashioned systems and beliefs and, as the account of a man trying to break into a conversation with two senior women indicates, the dynamics are changing as more women feed into the system and progress to the highest levels.

Social science studies provide ammunition for anyone who needs it, to spell out the environmental disadvantages that 'male by default' attitudes may continue to impose in some quarters and which continue to propagate.[43] The evidence will also help to inform what changes and interventions are needed, a topic I will explore at the end of the book. Before that, I want to move on to what happens when men and women are 'measured', as systems tend to require, for instance around appointments and promotion. Here too there are systemic issues, with stereotypes and different expectations apparent, even when nominally objective metrics are used: many of the criteria used to judge scientists are, in practice, inherently biased.

CHAPTER 7

GENDERED SLINGS AND ARROWS

Men can have careers and marriages, so why can't women?

Jocelyn Bell Burnell[1]

Discrimination isn't a thunderbolt, it isn't an abrupt slap in the face. It's the slow drumbeat of being unappreciated, feeling uncomfortable, and encountering roadblocks along the path to success.

Meg Urry[2]

She's a woman. Women do not run their own lab; they work in the lab of their husband.

unnamed scientist as quoted by Tom Steitz to Venki Ramakrishnan[3]

The Leaky Pipeline

As in any professional setting, in STEM careers women thin out as one moves up the seniority ladder. Some of the reasons for this loss remain sadly universal to professional life; some are more specific to STEM. It is easy to look at the numbers of women doing STEM subjects at university and think this problem is over and done with. Women are indeed entering university to study these subjects in large numbers, but with huge variations between disciplines: at Cambridge University 80% of undergraduate vets are women, and women comprise around 50% of chemists, but only around 20% of physicists. Moreover, women don't progress in line with these numbers. The so-called leaky pipeline continues to leak, as it has for many years, although one might argue that in many cases the women don't leak out, they are pushed, or at least encouraged to go by the environment in which they find themselves. The scissor graphs of the previous chapter illustrate the leak graphically. Some people object to the terminology of a 'leaky pipeline': some of the women dropping out of research, or the wider STEM arena, may be making a conscious and positive choice, just as many men do. The leaky pipeline concept may be inappropriate to apply to them. Nevertheless, the phrase remains the most commonly used and effective term to describe the many women for whom an unsupportive or even actively hostile environment forms a significant factor in their departure.

How fast women leak out of the scientific research career pipeline varies by discipline and country. Many qualified women don't only leak from academia, but from the entire scientific workforce, turning their back on their love for the subject and their invested years in education and training. It is very hard not to see this as a waste of talent. A 2022 report on the state of girls and women in STEM from the US National Girls Collaborative Project found that, whereas women obtained half the first degrees in Science and Engineering, they constituted only 34% of the STEM workforce, averaged across the disciplines.[4] The percentage was 48% in the life sciences, dropping to 26% in computer and mathematical sciences, with a further drop to just 16% in engineering. In the UK, the Women in Science and Engineering (WISE) campaign states that women made up 24.2% of the core STEM workforce in 2019–20, with this percentage having slowly risen over the years, but again the numbers vary across the sectors, with only 10.4% in engineering.[5]

A European Research Council funded project looked at the issue across seven different European countries (Switzerland, Belgium, the Netherlands, Austria, Iceland, Slovenia, and Italy), highlighting both the similarities and the differences in career trajectories.[6] Even at the PhD level, women were under-represented in Switzerland, Belgium, the Netherlands, and Austria, but Switzerland actually had 25% women in the full professor ranks, the highest of all the countries, while in Belgium and the Netherland this proportion remained below 15%. The story in each country will reflect cultural differences, making this a tricky problem to analyse let alone resolve, and once one takes into account disciplinary variations it becomes even more complex (the data in this study covers all subjects, but does highlight the, sometimes stark, differences between the STEM and other disciplines).

The critical stage at which the largest number of women leak from the career pipeline that they have already entered, is the one when women move from being a dependent researcher (graduate student or postdoc) to establishing themselves as fully independent and with a tenured position, meaning their long-term future is secured. The scissor graphs of Figures 3 and 4, show the dramatic fall-off in numbers at these points. This specific staging post seems to be the key stage to focus on, identifying which issues deter women in so much larger numbers than men, even once they are committed to a particular scientific discipline and career.

Being in a tiny minority can itself be demotivating, a problem which will become more acute as a researcher moves up the ladder due to the fall-off in numbers. The writer Eileen Pollack describes this sense of isolation vividly in her book, referring to the obstacles she faced as 'mainly psychological'.[7] Pollack, my approximate contemporary from the USA, trained originally as a physicist. She attended Yale just after they had first admitted women, something which happened only in 1969. She left the field for a career in writing, because of the loneliness and neglect she endured in her department. Unfortunately, in a subject such as physics, where numbers of women remain stubbornly low, that sense of isolation may still be felt by those few women. It can, as in Pollack's case, be a real deterrent to continuing with the subject.

It is undoubtedly the case that many early career researchers drop out: research going wrong is discouraging and permanent jobs are hard to find. Those facts are equally true for men and women, but the women may also be pressed to find support; may instead find themselves on the wrong side of biased professors and supervisors (and that doesn't only apply to men; women can be biased against women too).[8] They may be actively harassed, bullied or even assaulted by their research supervisors. Too often the expectation remains that women will bear the brunt of caring responsibilities.

The timing of wanting to start a family tends to be the time of precarity in an early researcher's career, before any sort of security of tenure has been achieved, exacerbating the challenges of parenting for women. Two to three years for a postdoc position is common, but often the posts are even shorter. Maternity leave may not be accessible until the individual has been in post some time. To this should be added the common belief (amongst promotion and appointment panels at least) that moving between labs improves your experience, introducing another potential complication for women: women with young families are likely to be less mobile in general, less able to take a job involving a weekly commute, say, or to move geographically between distant positions to progress their career, while taking their partners as the 'trailing' spouse.

Overall, there are many factors that can make it hard to balance progression with a settled life, having children is just one of them. For many this is too unattractive to make it worth pursuing and, not infrequently, it can lead to a woman moving out of academia into a sector where permanency of positions is easier to come by, thereby giving more certainty to their lives. Beyond this, there are the attitudes of those around you to contend with. As one (academic) woman put it to me in an email '*I am acutely aware that people tend to view you differently if they know you have children*'.

The Impact of Family

Historically, women were expected to be the care-givers. Before adequate birth control measures were readily available, women frequently spent years of their adult life either pregnant or (unless their society and financial means favoured wet-nurses) breastfeeding. During those years they are hardly likely to be at their physical best. In 17 years, from 1839–56, Charles Darwin—to take a scientist at random—fathered 10 children. That means his wife was pregnant for nearly half of that time (there may of course also have been early miscarriages which would have been both physically and mentally devastating). Emma Darwin (née Wedgewood) may not have had any aspirations to be a natural philosopher, but even had she wanted to follow that path, her physical strength might not have permitted it. Mary Somerville, as we saw earlier, only came to attention as a scientist in mid-life, after she had had her five children.

What was regarded as normal in Victorian England still seems to have overtones today. Women, marriage and children may still be seen as incompatible with being a top-flight scientist. Ottoline Leyser, a Cambridge academic and now head of UKRI, compiled a book she called *Mothers and Science*.[9] As she put it:

> *I put the book together because I was so frustrated with the prevailing pessimism that is endemic in science in general: that everything is so difficult and that anything you do that is a bit unusual immediately puts a black mark against you. Having children has somehow got itself included in that list of things that are impossible in a research career. And it's just not true. If you look around there are many, many people who have successfully combined those things. I sometimes say that it's perfectly possible to combine a career in science with children, because men have been doing it for centuries. I do think that is an important point. The part where you're actually giving birth is a relatively short part of the process. The real issue is managing your flexibility of time for the next 20 years, and that's not a gender issue—it's a parent issue.*[10]

However, the reality on the ground is that more women than men leave science as a result of becoming a parent. A 2019 study from the US showed that, whereas more than 40% of women with full-time jobs in science left the sector completely or went part-time after having their first child, only 23% of new fathers left or cut their working hours.[11] In the US, unusually in the developed world, there is no statutory maternity leave, although of course many employers do have their own maternity leave policies (which may be unpaid). Times have moved on since the conversation I overheard between two American women in the 1990s, planning to time their pregnancies to allow them to give birth over the summer break so that they would not have to take any unpaid leave (at the time I thought, good luck, pregnancies don't come along to order), but not as far in the US as in some countries.

These days the idea that professional women should take time out to have a child may be accepted, and allowances may be made, but this does not mean there are not multiple ways in which mothers may be disadvantaged. Stopping the tenure clock in the US[12] is normal there for parents, both male and female. But that will not take the pressure off a parent. It does not necessarily mean they can relax and enjoy the baby. Furthermore, the policy may be superficially gender neutral, but a study (in this case of economists) showed that the adoption of tenure clock stopping policies substantially reduced female tenure rates, while simultaneously increasing male tenure rates.[13] Not, one presumes, the intended outcome.

Attitudes towards becoming a parent exhibit something of a double standard, as Meg Urry, the first woman to be tenured in Yale's Physics department, pointed out:

> *When I told my thesis adviser I was pregnant, he said, 'So, you want to have it all!' I smiled but later thought, wait a minute, isn't that what all you guys have? Why is it 'all' for me and 'normal' for you?*[14]

Fathers were never regarded as unusual in the physics—or any other—lab, but for a woman to want to 'have it all' was regarded as very odd indeed.

Anne McLaren, the first woman to become a Vice President of the Royal Society, said of her young parenting years more than fifty years ago:

> *When I was young, I never thought of myself as a woman scientist, just as a scientist, and as a woman. There was no statutory maternity leave, we just had children and got on with things as best we could.*[15]

For me, when I had children, it was rather similar. There was some maternity leave—I got sixteen weeks on full pay, unusually generous for the time—but there was no option to work part-time or flexible hours. When I asked for two weeks unpaid leave after those sixteen weeks with my second child, it caused a bit of consternation. I was told '*no academic had asked for that before*', but it was granted. Thereafter, my husband and I shared the childcare, alongside part-time nursery places because that was all we could get; nurseries were at the time still a very rare phenomenon in the UK. Many academic institutions, but by no means all, can be (invisibly) flexible, only caring about getting the work done rather than presenteeism amongst its faculty. I certainly used that flexibility to the full, working decidedly non-standard hours around the nursery care. Not everyone can. Not everyone has the kind of supportive husband I had, who not only took his full whack of child-care, meeting the children from school and going above and beyond, but let his own paid career falter and then fold as a result.

I worry that these days it is harder to be flexible in the way that McLaren and I managed. The option for a parent to declare a desire to work part-time, or asking for flexible hours to do the school run, can all too easily be translated into 'this scientist is not serious', with a consequent perception penalty by others around them. It should not be so, but unfortunately too often it is. McLaren declared '*As a research scientist the hours are flexible. When the babies were very young, I used to take them into the lab*', which you'd never get away with on safety grounds now. It is both easier today, because there are formal routes to state what an individual wants to do to address their caring responsibilities, and harder, because such declarations may become an invisible stick with which to beat that individual.

For some women, if the work environment is sufficiently unpleasant, the 'excuse' of children may be a face-saving way of extricating themselves from it. There have been attempts to examine how productivity is affected by motherhood, with the usual assumption, backed up by some studies, that there is bound to be a negative impact. However, when looked at in a nuanced way it is obvious that how parental responsibilities are split between a dual-career family has significant consequences, as well as the age and number of children at the time under study.[16,17] These are tricky arguments to try to unpick and the debate can provoke strongly polarized opinions.

Problems may also arise due to family life when it comes to some of the add-ons that form a significant part of an academic's life, directly impacting on their promotion prospects. These include the desirability of travel—to conferences, field work, committee meetings and for collaborations—and the willingness to work far more than a formally-agreed working week. Such demands don't sit well with caring responsibilities, whatever your gender. However, as the building evidence around the impact of Covid-19 highlights, women do seem to bear a disproportionate amount of the burden of caring, whether for children or for elderly or sick relatives.[18] This can add up to intolerable stresses which, if other parts of someone's career are not going well, may simply tip the balance into that individual walking away for some less demanding job. Nevertheless, there also seems to be a sense that the pandemic has meant that many more men have had to face up to the challenges of childcare and it is easier to discuss. One woman said privately to me that men ... '*are far* FAR [sic] *more open about the impact being a parent has on their work than they had been previously.*' She concluded that '*the conversation was opening up in a positive way*'. Time will tell if this opening up of the conversation around parenting persists in the long-term.

Some group heads expect their teams to work extraordinarily long hours, turning up early in the morning themselves, or lurking late at night to check their students are still beavering away at the bench. There is the presumption that more is better, despite the evidence that beyond a certain point (economists seem to put this at about 50 hours a week), it is a case of diminishing returns: longer does not mean more, but instead leads to many other problems, including mental health issues and a drop in the ability to relate well to others. Of course, in many experimental areas, there may be periods when the demands of an experiment do require attendance in the small hours to check how the experiment is evolving or to change a sample's environment, but these periods should be both rare and short in duration. Certainly not enough to disrupt family life fundamentally, however inconvenient the night shift may be in the short term. Furthermore, those people judging outputs when it comes to appointments or promotion should be able to factor in productivity as it is impacted by, for instance, pregnancy and maternity leave or chronic health issues. Brilliance is not simply determined by weight of papers; the quality of papers matters too.

Bias

In judging individuals, it might be thought that there are appropriate quantitative and objective metrics to be used. In reality, such metrics can be seen to disadvantage women. Different disciplines and different countries may exhibit these tendencies to a greater or lesser extent. As we will see, letters of reference—in science as elsewhere—tend to use fewer stand-out adjectives about women than men, meaning their chance of progression is reduced. Women's papers are cited less;[19] their grants are on average smaller;[20] and their papers have a harder time getting past reviewers.[21]

A recent study of referees' comments highlighted just how unpleasant, not to mention unhelpful, referee comments may be.[22] One example stated: '*This paper is, simply, manure*'. Hardly constructive criticism. And misogyny can feed into reviewers' comments, sometimes explicitly, as in the case of another review quoted in the same paper) '*The first author was a woman. She should be in the kitchen, not writing papers*'. I would like to think that referee was blacklisted thereafter by the editor concerned, but the fact the editor saw fit to pass the comments on makes me think that was unlikely. Faced with such responses many researchers' confidence, and particularly those of women who are in the minority in a field, may be so shaken that they step back or quit altogether, contributing to the 'leaks' from the pipeline.

Underpinning many of the obstacles I've just outlined is bias, unconscious though it may be. Overt discrimination is not only illegal, it is by now less evident than in the quote from Venki at the start of this chapter. Rita Colwell, the bacteriologist and first woman to head up the US's major funding agency, the National

Science Foundation (NSF) was told at the start of her career in 1956 that '*We don't waste fellowships on women.*'[23] This remark made her angry, but it didn't stop her in her professional tracks, as it might have for many women then and since. Nevertheless, most senior academics would probably be more cautious to express such an opinion outright today, whether or not they privately harboured such thoughts.

But it is not necessary to be aware of thinking that women are somehow second-class citizens; that opinion may seep into actions unconsciously. Bias of this subterranean sort—variously known as unconscious or implicit bias—has come under intense scrutiny in recent years, as it should. Bias, at the individual level (as opposed to the systemic kinds I identified above), acts as an unconscious reaction to all the stereotypes we have been fed since birth, and comes in many shapes and forms. It can be compared with (Economics) Nobel Prize-winning Israeli psychologist Daniel Kahneman's two modes of thinking, System I and System 2.[24] System 1, which operates automatically and quickly, in an essentially involuntary way, would give rise to unconscious bias of the sort that decides a woman is not as capable of being a scientist as a man but would be expected to be very good at childminding or nurturing more generally. System 2, the slower thinking process, takes the time to think through such a decision. That process allows the bias against women in that first thought to be teased out, confronted and, hopefully, rejected.

Organizations that introduce unconscious bias training need to ensure that the need for moving on to slower, more considered (i.e. System 2) thinking is impressed on the individual, not imagine that the training is simply some sort of tickbox exercise telling people they should not be biased. The latter, too commonly seen in online courses in particular, is totally insufficient to see outcomes shift, the only measure of the success of such a programme. It may even backfire.[25] The subtle ways in which unconscious bias can operate in an academic setting was spelled out at length in Virginia Valian's classic 1999 book *Why So Slow?*[26] about the progression of women in academia. Despite Valian's book being a quarter of a century old, it is still a sobering read, highlighting all the different places where disadvantage may accrue across the university sector (and not just STEM).

What is the hard evidence, beyond anecdote and suspicion, that unconscious bias impacts on women's careers? Increasing numbers of studies show, in many different guises, just how potent such bias can be. One of the most striking classes of study is that which compares the reactions of both men and women to identical CVs submitted under a typically male and a typically female name. Valian highlights a study from as far back as 1975 by L.S. Fidell which demonstrates bias in her own field of psychology.[27] Many studies since have gone on to demonstrate the pervasiveness of such bias, which does not seem to be disappearing. For instance, in one much-cited study, faculty were sent identical CVs to evaluate, differing only in whether the name at the top appeared to be male or female.[28] These were application materials for an undergraduate science student who had ostensibly applied for a science laboratory manager position. Both male and female faculty were more likely to 'hire' the man, as well as offer him more support/training and a higher salary than the woman, despite the identical track records.

The late 1990s saw not only the publication of Valian's book, but also a paper regarding data from the Swedish Medical Research

Council concerning biomedical fellowships, which used a Freedom of Information request to obtain the actual evaluation sheets used by the peer-review panel.[29] Sweden, it should be noted, is generally regarded as one of the most egalitarian societies in the world when it comes to gender issues, but the findings would not have led a reader make that assessment. The title of the paper presenting the results of this study gives the game away: 'Nepotism and Sexism in Peer-Review'. Using an array of metrics to devise a figure of merit for impact, it demonstrated graphically, as shown in Figure 6, how great a difference there was in evaluators' competence scores for men and women objectively assessed to have demonstrated equivalent impact.

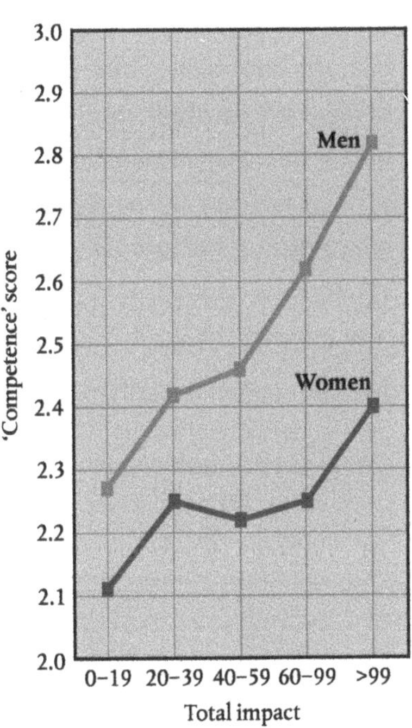

Figure 6 Mean competence score given to male and female applicants by the MRC reviewers as a function of their scientific productivity, measured as total impact.
Christine Wennerås and Agnes Wold, 'Nepotism and sexism in peer-review', *Nature* 387 (1997): pp. 341–3, doi: 10.1038/387341a0. Used with permission.

The authors' analysis showed that this discrepancy amounted to the equivalent of three papers in a high-ranking journal, such as *Nature*, or the phenomenal difference of 20 papers in a moderately highly ranked journal. Equally worrying was personal bias, the nepotism of the title, when an applicant was known to a panel member. Despite that particular person not being allowed to take part in the evaluation itself, as is customary with such funding panels, it transpired that the other panel members scored anyone known to have such an association more highly. Bias comes in many forms, however much processes, such as excluding a known associate from making the relevant judgement, may attempt to overcome them.

When this study appeared, back in 1997 many women will already have had suspicions that they were being treated unfairly, or at least differentially compared with men. For far too many of us, the message of the 1988 'Miss Triggs' cartoon (Figure 7) will have felt painfully familiar then, and may still do so now. To counter this sort of behaviour it is important for those around the table to chip in, to remind everyone that Miss Triggs did just say this and it is good that Mr X agrees. This technique was brought more forcefully into the public's eyes by female staffers in the Obama White House, who called it amplification. As the Washington Post described it:

> *When a woman made a key point, other women would repeat it, giving credit to its author. This forced the men in the room to recognize the contribution—and denied them the chance to claim the idea as their own.*[30]

Every committee around the world, in academia or not, could do with more pushing back on bad behaviour from other committee members. The other tendency seen only too often at committees,

Figure 7 1988 *Punch* cartoon by Riana Duncan, 8 January 1988. Used with permission.

debates, and other potentially confrontational situations, is for men to talk over another committee member typically, but not necessarily, a woman. Again, American politics shows a clear example of how to deal with this arrogant behaviour; in the 2020 Vice Presidential debate, Kamala Harris calmly said over and over, '*Mr Vice President, I'm speaking.*'[31] Many people would, however, find it easier if someone else, an ally male or female, made that very same point for them.

Systemic Bias

Whatever women may feel in a given situation, evidence of bias is very different from suspicion. Nancy Hopkins, a distinguished biologist at MIT, may initially have believed her comparative lack of success in her career was down to her own shortcomings, but once she sat down to look at the data, she realized this was not so. Motivated to look into this when her department head refused her request for additional space (for fish tanks to house the zebra fish she was keen to start researching) she started to get bitter and to wonder whether there was something more behind her inability to progress as fast as her male colleagues; explicitly she wondered whether her lab space was in fact smaller than her male colleagues. To test this idea, she got out a tape measure and went round all the labs to work out how much space the men and women in her faculty actually had. Realizing, when she had collated her results, that there was systematic disadvantage for the women, and remembering all the other situations in which she had found herself that seemed to imply discrimination, she drafted a letter to the then President of MIT spelling the situation out. As she describes it:

> *I said, you know, there's a terrible chronic problem in this institution. I'm sure you don't know about it, because if you did, you'd certainly want to fix it. I described some aspects of what I thought was discrimination against women, preventing them from doing science the way they should be able to.*[32]

Showing this letter to the 15 other women in her School of Science before sending it off, it transpired they were all keen to sign too.

Following on from the letter she sent to President Vest, which provoked a strong and immediate response, a very significant piece of work was undertaken by MIT, leading to a full report, *A Study on the Status of Women Faculty in Science at MIT*, being published in 1999 analysing the situation.[33] By way of introduction to the Report, Vest said:

> *I learned two particularly important lessons from this report and from discussions while it was being crafted. First, I have always believed that contemporary gender discrimination within universities is part reality and part perception. True, but I now understand that reality is by far the greater part of the balance. Second, I, like most of my male colleagues, believe that we are highly supportive of our junior women faculty members. This also is true. They generally are content and well supported in many, though not all dimensions. However, I sat bolt upright in my chair when a senior woman, who has felt unfairly treated for some time, said 'I also felt very positive when I was young.'*

Key to the report was the evidence gathered demonstrating how women were systematically disadvantaged, including through marginalization and exclusion from important decision-making bodies:

> *Examination of data revealed that marginalization was often accompanied by differences in salary, space, awards, resources, and response to outside offers between men and women faculty with women receiving less despite professional accomplishments equal to those of their male colleagues.*

It is important to note that last point; these women were remarkable indeed, with four out of the 16 having been awarded the National Medal of Science, compared with seven of the 162 male full professors in the science departments. Additionally, 11 of them would go on to be elected members of the National Academies of Sciences, Engineering, and Medicine, compared to only 11 in total of the 162 male professors. The report showed that disadvantage and discrimination were systemic, despite the women's stellar attributes. This was not a case of affirmative action leading to weak appointments, nor women just not being up to the mark of their male colleagues. It is no wonder that Vest sat bolt upright. He was right to do so, and it was something of a wake-up call for many around the world.

I first came across this Report soon after its publication, when it was brought to my attention by a (male) professor at MIT who asked me if its findings tallied with my own experiences. At that time, I was already a professor at Cambridge, and had just been elected to the Royal Society, but I hadn't really given much thought to my gender. I had simply kept my head down and got on with the job, thinking things that didn't feel comfortable or 'right' were due to my own failings, not systemic issues. Exactly as Nancy Hopkins had thought at first. To some extent, the MIT report sensitized me to my situation. For that I was not particularly grateful at the time. I realized, just as Hopkins had, just as I'm sure many women did at the time and no doubt still do, that not all of my problems were down to my own weaknesses. I had reached a standing where I might have been expected to be better integrated into the decision-making process in my own department than I seemed to be, and I hadn't fully appreciated this. The view, expressed in the MIT Report, that:

> *In contrast to junior women, many tenured women faculty feel marginalized and excluded from a significant role in their departments. Marginalization increases as women progress through their careers at MIT.*

resonated with me.

I wrote a few years later (unpublished text, but it reflects my mood) about this time in an internal document,

> *I frequently felt thwarted and outmanoeuvred in strategic and political discussions, and I became very unhappy, seriously thinking about quitting Cambridge to move somewhere else where perhaps I could have fitted in better. It*

> *was never clear to me whether the issue was my character or my gender. It was easy to believe I did not fit in as my thought processes did not always mesh with my colleagues, and undoubtedly I did not always act in a cool and calm way. Although* [my head of department] *always appeared very supportive in his words, I did not believe he always meant it and sometimes felt patronized rather than respected. I suspect my election to the Royal Society in 1999 probably only exacerbated some of these issues.*

It was as a direct consequence of this frustration at the ingrained biases in my workplace that I took up the mantle of championing gender issues, in all their dimensions.

It is important to remember that, although women leave in far greater numbers than men at early stages in their career trajectory, this does not mean mid-career is without substantial challenges for women. By this point women may feel there is nowhere for them to go, so they don't 'leak', but they may also find themselves held back, loaded with excessive administration and without the funding they deserve. However, hurdles for women may arise at any stage in their career in ways that many men will not have to confront.

The MIT Report was published more than 20 years ago, but many women will still believe its message of systematic disadvantage applies today. This is far from a case of all men knowingly disadvantaging women. Were that the case it might be easier to handle. The problem arises from, in many cases, an unintentional unconscious bias leading to bad outcomes. Let me pause here a moment to say 'not all men ...'. There are many men who are champions for gender equality. There are many men—my husband notably amongst them—who share domestic responsibilities, including childcare, a factor that must not be overlooked. But, in the workplace, as so many studies have shown over so many years, stereotyping and assumptions about women's capabilities and appropriate roles, continue to lead to bias.

In 2011, MIT prepared a follow-up report to explore how the situation for women faculty had changed since the initial study.[34] During the interim, the percentage of women on the faculty had nearly doubled, but that in itself had led to problems of perception, with the worry that standards for hiring and promotion of women faculty were lower than for male faculty. This pernicious perception is sufficiently intangible that it is hard to counter, yet can linger in people's minds and damage the self-confidence of recent hires, as summed up by one woman quoted in the report *'I felt I was invited to interview because I was dazzling, but now I wonder ...'*. A further issue related to 'expected behaviour', illustrating another trap for women: *'There is an expectation of niceness, sweetness. It's everywhere. Students, collaborators all make this mistake.'*

Women who act nice may well not get very far. But actions seen as inappropriately aggressive in women may also be counterproductive. Not all the women were happy at being expected to be willing to talk about work–life balance or be a mentor; some did, but others saw this as equally unwelcome stereotyping.

As time has progressed, many scholarly studies have been carried out continuing to use fictitious CVs to explore different manifestations of bias. The 2012 study that I mentioned earlier is one such.[35] A subsequent (2019) study led by Asia Eaton of Florida International University, extended the range of stereotypes being considered in the case of hiring a postdoctoral researcher.[36] Again, identical CVs were submitted under a range of names, this time implying not only their gender but also ethnic background. The study found that women were disadvantaged over men, a situation compounded if they were perceived as non-white, though with different attitudes to ethnicities and stronger gender bias shown in physics compared to biology. There were, however, subtle differences in different fields (specifically physics and biology were compared). The study reported that:

> *Faculty in physics exhibited a gender bias favoring the male candidates as more competent and more hirable than the otherwise identical female candidates. Further, physics faculty rated Asian and White candidates as more competent and hirable than Black and Latinx candidates, while those in biology rated Asian candidates as more competent and hirable than Black candidates, and as more hireable than Latinx candidates.*

It seems, then, according to this study, that Physics faculty show not only a gender bias but bias against some ethnic minorities too.

Furthermore, when it came to intersectionality, which can compound the likelihood of bias but is a much less studied area, the same authors stated that:

> *An interaction between candidate gender and race emerged for those in physics, whereby Black women and Latinx women and men candidates were rated the lowest in hireability compared to all others. Women were rated more likeable than men candidates across departments.*

Being likeable, it should be noted, is not necessarily regarded as a strength in a postdoc, a point I will return to below. This study, therefore, highlights the problems of intersectionality; compounding factors such as these should never be forgotten, but their complexity makes them challenging to study.

Reference Letters

These studies, and others like them, demonstrate that, however unintentionally, gender bias is prone to turn up in hiring decisions, even if no other factors come into play. But, of course, they do. Academic hiring involves significant weight being attached to letters of reference which can be remarkably frank. I have seen some outrageous statements in my time, such as the letter of recommendation (or not) saying Professor XX had only got as far as she had because there were no other women in the field, and also that she was 'feisty', expressed with negative connotations. The committee I sat on, I am pleased to say, rejected that letter completely from consideration.

When it comes to gender, much has been written[37,38,39] about how different words are typically used about men and women, regardless of the gender of the person doing the commenting. Feisty, for instance, is not often applied to men and does not come across as intended to be a compliment when applied to women. The evidence from these papers suggests that men are more likely to be described by so-called 'stand-out' and 'agentic' words, women by grindstone and more passive adjectives. In recent years this has become such a known issue that there are even websites devoted to highlighting how gendered a letter is (the idea being to check your writing before it's sent off).[40] If a candidate for a job is described as 'hard-working', it may mean literally that and that no more can be said. It is, of course, not a rude thing to say. But if, in fact, the author is trying to convey that the person concerned is outstanding, deeply analytical or brilliant they should say so. Unfortunately, the evidence cited above is firmly that a woman is more likely to be described, quite unconsciously, by the first somewhat mild term, a man by the latter stand-out adjectives. Other attributes often used in a positive but not always helpful way about women may include being a good team worker, good at pastoral care and conscientious. And, as noted above, women may be described as likeable, another adjective that is not really relevant to the task of running a lab or a piece of equipment, or being gifted as a scientist, the sorts of skills one might hope for in a new hire in STEM.

Finally, there is the danger that men are judged on potential, given the benefit of the doubt, while women need to have proved themselves before they are able to be judged properly. Wolfgang von Goethe wrote '*Girls we love for what they are; young men for what they promise to be*' about two hundred years ago, albeit not in the same context, but it seems that little has changed. I once sat and watched two applicants being considered for promotion in an applied discipline. Both had patents to their name. When the woman was assessed, the question was asked if the patents had translated into a product and profit; no such question was asked about the man. One of the other committee members, a man, spotted this double standard for the problem it was, and the issue (in the case of the woman) was dropped, with the recognition that there had to be the same assessment criteria for both. It would, in my view, have been equally appropriate to have asked about the success of the man's patents.

Using Metrics as Criteria

It might be anticipated that some metrics in an academic's life are so objective that there cannot be any opportunity for bias to creep in. One such figure of merit might be expected to be citations—how often someone's publication is quoted in other papers. But even this turns out to be curiously gendered. Personally, if I'm reading a paper, I don't look to see who the lead author is and whether they are male and female. I may of course be subliminally affected, and if it is by someone I know, that will naturally get factored in according to how highly I rate them. However, it has been shown statistically that men are more likely both to cite their own work (in indices of citation who is doing the citing is not always examined, consequently often heavy self-citations can pass unnoticed) plus other men's work more than that of women.

One recent study, extending over decades, showed that men cited their own papers 56% more than women.[41] Women, on the other hand, were shown to be likely not to cite their own work at all, a case of inappropriate modesty which may damage their careers in a rather direct way. Furthermore, when analysis is carried out focussing on the key positions in an author list (sole author, or first and last names in a list where names are not listed alphabetically as they are in some disciplines), it has been demonstrated that when a woman's name was listed in any of these positions, the paper garnered fewer citations than when a man's name appeared in these roles.[42]

A study looking specifically at the references in papers published in five key journals in neuroscience, illustrated this graphically (Figure 8).[43] The figure compares the gender composition that had men as both first and last authors (MM), relative to those within papers that had women as either first or last authors, referred to as W∪W and comprising WM, MW and WW,[44] according to who did the citing. Figure 8 shows that, by comparison with a statistically derived norm, papers published by two men (MM) over-cite other papers written by MM. As the authors suggest, this may derive directly from some internal belief that a woman's work cannot be as good as a man's, but also systemic bias.

Thus, the evidence is clear on the female citation deficit, although there may well be disciplinary variations as was the

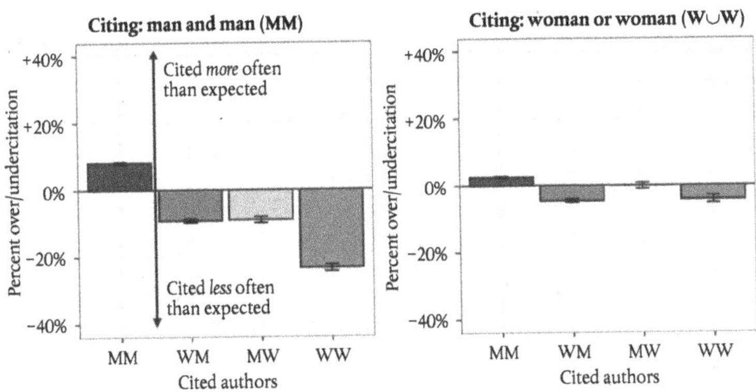

Figure 8 Degree of over- and under-citation of different author genders within MM and W∪W reference lists.

Jordan Dworkin, Kristin Linn, Erin Teich, Perry Zurn, Russell Shinohara, and Danielle Bassett, 'The extent and drivers of gender imbalance in neuroscience reference lists.' *Nature Neuroscience* 23 (2020): pp. 918–26 doi: 10.1038/s41593-020-0658-y. Used with permission.

case with bias in reading CVs. In the case of citations, part of these differences may arise from who knows whom, whom you network with and hence whose work you may be most familiar with. Since networking patterns indicate that men tend to collaborate more with other men than with women, this would seem to provide a reason for the lower numbers of female citations, but it is a young area of research specifically exploring connections between these two different observations, to look for causation rather than simply correlation.

Similar gendering and bias have been found in many of the other 'figures of merit' associated with academic progression. Although a recent study from the US showed that, as measured by National Institute of Health funding, so it refers specifically to life scientists, once a woman had begun to be established as an independent researcher the likelihood of her receiving future grants was essentially the same as a man's, not all analyses convey the same rosy picture. A 2020 analysis of the size of awards made to men and women Principal Investigators by UK Research and Innovation (UKRI), the UK's umbrella funding organization covering all disciplines, showed that the median award value for female awardees was approximately 15% less than the median award values for males.[45] The European Research Council (ERC) has consistently monitored the gender split of awards they make and the relative success rates. Initially there were some very significant differences in success but, across the disciplines, the levels are now more or less equal after extensive consideration of the ERC processes to try to eradicate bias.

Although it may feel like a quick fix simply to increase the number of women on decision-making panels to reach some pre-determined level (a quota), for all the reasons I've given earlier in this chapter, this is unlikely to resolve the issue: women can be just as biased as men when it comes to judging other women, having internalized lower expectations of their gender. I overlapped as an ERC Scientific Council member with Isabelle Vernos, a French biologist working in Spain, who wrote, during her time as Chair of their Gender Balance Working Group,

> *Quotas might even make matters worse by overworking already-stretched female scientists. Instead, a range of bottom-up and top-down measures are needed to effect lasting change in the structures and culture of science.*[46]

There is no doubt that women, as with other minorities, get a wide range of what has been termed 'housekeeping' dumped on them. This too is a burden, as Vernos makes clear. If a committee is required to have a quota, say 40%, of women then, in a subject such as my own (Physics), where typically there will be 20% or fewer women on the faculty, it inevitably means those women end up doing twice as much work as the typical man. If the committee is of great importance, well and good, but unfortunately women are more likely to be stuck on those committees with less influence, a fact that is true in many sectors. Even on grant-giving committees, which are undoubtedly important (including for networking but also influence), nevertheless the mental and time burden may more than offset the advantages. Yet women are not good at saying no.

The German funder, the DFG (Deutsche Forschungsgemeinschaft), in 2020 made a statement addressing these problems, indicating that the minimum proportion of women on a committee will be based on the current percentage of women eligible to sit on that committee.[47] They are also pushing to give more recognition to committee work in assessing everyone for recruitment and progression. It does seem only reasonable that 'service' to the academic community should be valued, but that is not always the case in practice.

Thus, we see that letters of reference, citations, and size of grants all indicate a gender difference, always to the woman's disadvantage. *The Metric Tide* report, produced to assist UK funders in assessing the reliability of different types of purely quantitative assessment criteria, cited many of the papers regarding bias I've quoted here. Worried by the likelihood of metrics used crudely in perpetuating inequalities, their conclusion was that: '*.... issues with the use of quantitative indicators to assess research outputs include concerns over gender bias in citation practices and resultant effects on equality and diversity*'.[48]

Some of these issues only apply to the academic scientist. But unconscious bias (for instance in letters of reference, where these are still used beyond the education sector) applies across the board. It is easy to think of ways in which bias can creep in unconsciously and scupper the woman's chance or hold her back so that at later career stages she has not achieved her full potential or, indeed, the salary or standing of her equivalent male colleagues. Many of these issues are not specific to STEM or academia, but they all make a huge impact on the women who do enter the STEM pipeline.

Networking

There is no doubt that the ability to network, to build relationships through meetings formal and informal, can play a significant role in a scientist's career. This is why attending conferences can be of such significance, particularly in the early part of a career where the scientist is wishing to penetrate existing networks. The socializing that occurs in the margins can be a wonderful opportunity to approach others, be they higher or lower in the pecking order, and to drill down into technical detail, to propose future joint projects, or simply to put a face to a name. Networking is, however, one of those areas where minorities can be at a disadvantage simply because of social norms and practice. Going out for a drink after work is a long-established habit in which many sectors indulge. However, by deliberate exclusion, the need to be at home, religious background or social awkwardness, many do not find it a habit in which it is either easy or comfortable to take part. Women may be anxious about suggesting to a man a desire to head off to a bar to talk things over; the man might be anxious about accepting such an invitation. A man, on the other hand might not think twice about either making or accepting such an invitation from another man.

Networking also favours the confident. Individuals may be shy, whatever their gender. If a researcher is suffering from impostor syndrome, they are much less likely to plunge into the unknown by approaching a hot-shot professor and more likely to lurk around the fringes of the room. Without a mentor to facilitate an introduction, the more retiring will be condemned to losing such a potentially productive opportunity.

Covid-19 has been somewhat democratizing, at least in the short term. The remote videoconference meetings necessarily introduced during the Covid-19 pandemic, has equally disadvantaged all of us when it comes to networking. But it is difficult to build up a relationship with someone you have never met, whose body language you can only read via a small virtual rectangle and whose facial expressions may be obscured by the way the light falls on their features. However, maybe we should all learn to do this better. Not travelling halfway across the world to present a half hour talk not only means less time spent travelling but reduces a scientist's carbon footprint. There may be other advantages in this transformation of how networking opportunities and the construction of collaborations can be carried out. It could remove the unfavourable comparison of those who won't or can't travel, because of caring responsibilities or for any other reason, with those who are perfectly content to spend a lot of their life jetting between meetings. It should anyhow be questioned how good turning up at a conference thousands of miles from home is as a discriminator between job applicants. My university has moved to a position where it at least values *invitations* as much as their delivery at serious international conferences, so that those who are approached to give significant talks, but decline for personal reasons or other professional commitments, still get the kudos of the invitation.

Harassment and Belittling

However important conferences may be, socializing at conferences brings its own hazard. I have not, myself, ever been formally propositioned by a drunk colleague, but I have had some very unpleasant experiences of a predatory nature. Many women (and some men) fare much worse. One scenario occurs when women, seeking faculty positions, get invited up to someone's room to 'discuss the opportunity further' which turns into something much less professional and advantageous. Others may involve lower-level groping or harassment, carried out in public view but in a way that no one thinks to intervene. I've had leering colleagues do that, pinning me to the wall and making offensive remarks but in such a way that it is hard to escape. Why are women brought up not 'to make a fuss' or not to slap someone in full view of others?

Harassment may take many forms, verbal and physical, and it certainly does not need to be sexual in nature even when it is gendered. Many women may just be belittled, publicly or privately. Shirley Chiang, a nanophysicist from UC Davis, tells the story of how a bigshot professor opened a conference with '*Gentlemen and Shirley* ...'. Meg Urry tells a similar story about being singled out when the only woman in a lecture theatre. She also tells of a different example of the belittlement of women in full view:

> *When I was a young astrophysics postdoc at MIT (and the only female postdoc), one weekly colloquium speaker began his talk about the importance of high resolution in optical imaging with a badly out-of-focus slide. As he*

> *sharpened the focus to make his point, a topless woman in a grass skirt on a Hawaiian beach gradually appeared. The male students laughed, while the one other woman in the room shared an appalled look with me before standing up and walking out. No one ever told this speaker that his choice of slide was inappropriate. I intended to talk to him afterward, but I left the talk after about 20 minutes, having realized that I hadn't heard a word he'd said.* [49]

Predation does occur, but as a wide-ranging report from the US National Academies of Science[50] graphically illustrated (see Figure 9), much of the gendered harassment that does occur is of this denigration and belittling variety.

US anthropologist Kathryn Clancy described it in the following way, when presenting to a Congressional hearing in 2018:

> *Sexual harassment comes in two main forms—come ons, which are unwanted sexual advances and sexual coercion, and put downs, also called gender harassment, non-sexual behaviors that are crude or hostile regarding gender. While the come ons are the types of behaviors you see in articles about Harvey Weinstein and in sexual harassment trainings, the majority of sexual harassment are in fact the put downs.*[51]

However, this sort of harassment is quite enough to drive those who suffer it out of science. The National Academies report interviewed victims who described how they had skipped professional meetings and social situations, given up on research projects, and left jobs, just to avoid harassment. The emotions they described included being mortified, devastated, even outraged in some cases. But, as in so many spheres, many never formally reported this harassment. This might be because of the fear of retaliation, but also because of the depressing fact, true across all sectors, that reporting can lead to long drawn-out proceedings, taking

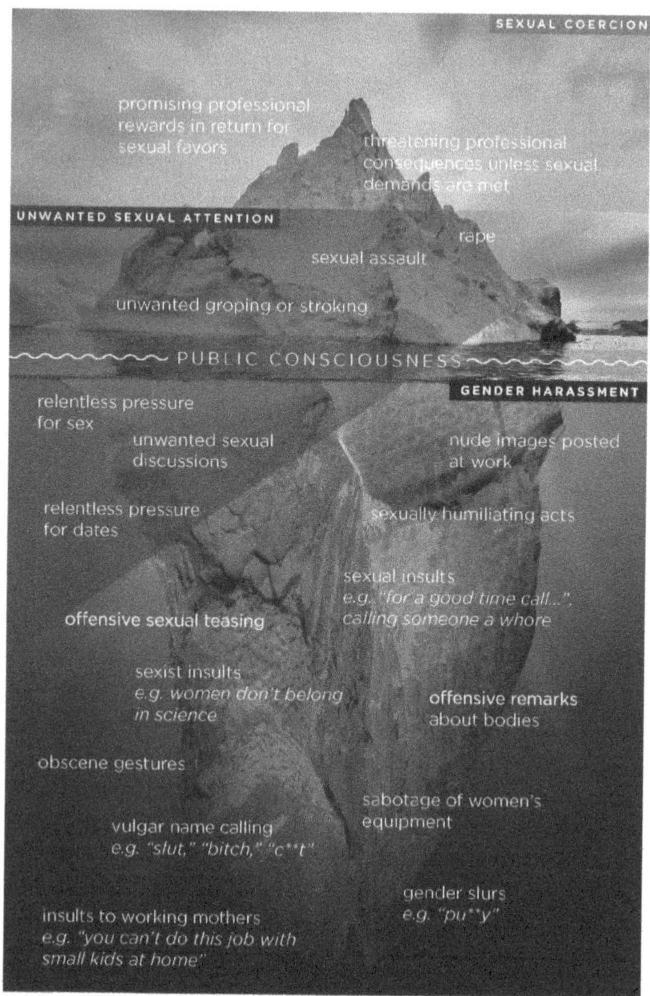

Figure 9 The public consciousness of sexual harassment and specific sexually harassing behaviours, much of which is 'below the water line'. National Academies of Sciences, Engineering, and Medicine, 2018. *Sexual Harassment of Women: Climate, Culture, and Consequences in Academic Sciences, Engineering, and Medicine.* Washington, DC: The National Academies Press. https://doi.org/10.17226/24994. Used with permission.

up precious time and energy, but often without ever reaching a satisfactory outcome.

The power of hierarchies in science, particularly for those just setting out on their careers, means that for many junior staff and students the fear of retaliation is only too real, even in situations where no formal complaint has been made. That retaliation does not need to be anything very visible. For instance, those precious letters of reference I discussed previously can simply contain a phrase such as troublemaker, difficult or—at an extreme—a liar. Or, if asked to comment on the applicant's work, they can downplay its significance, underrating it and thereby leading to a lower opinion of it than it deserves.

I once hypothesized on my personal blog that those who behave improperly and harass others fit into one of three categories: those who lack all empathy and cannot put themselves into the other person's shoes; those whose upbringing has given them such a sense of superiority they believe they are entitled to treat those they deem to be lesser mortals like dirt; and those whose own sense of self-worth is so shaky they can only believe in themselves if they believe those around them are insignificant. This is not an evidence-based conclusion, there may well be other factors that come into play. However, the case of an aggrieved male taking aim at female physicists was well illustrated by the public lecture Alessandro Strumia gave at CERN in 2018 to an audience at a conference on High Energy Physics and Gender. His remarks were regarded by CERN as 'highly offensive' and breaching their core values, leading them to decide not to extend Professor Strumia's status of Guest Professor.[52] He claimed, for instance, that women in the field were being hired over better-qualified men, stating that Physics was '*invented and built by men, it's not by invitation*', and highlighting the number of citations as an indicator of excellence.[53] (I have already highlighted the bias in citation rates; they are not a good measure.) It subsequently transpired that he had lost out on a job to a woman, a woman who was actually in the audience at the time.

Personally, I find it surprising that harassment, broadly defined, can occur at any stage in life. It is not just an early career problem. It astonishes me that, even as a senior professor, I have still come up against unsuitable behaviour, including of the sexually inappropriate variety. STEM has its own version of the #MeToo movement, simply known as #MeTooSTEM. Just as with the Harvey Weinstein case in the movie industry, because of the issues around power differentials in academia, serial predators can continue their activities beneath the radar for a long time, possibly a decade or more.

In the case of UC Berkeley's highly regarded astrophysicist Geoff Marcy, rumours were circulating for years without formal complaints being filed. Such an absence will always make it hard to investigate, let alone to prove, any wrong-doing. When many victims speak up, though, it is much harder for any cover-up or wilful blindness to continue. Ultimately, the wide range of voices speaking out against Marcy led to his fall from grace and he resigned from UC Berkeley without a formal disciplinary process taking place, but with the University finding that '*Professor Geoff Marcy violated the University's sexual harassment policies*'.[54] This situation was then followed up with his expulsion from the National Academies of Science, the first time such an act had taken place following changes to their by-laws permitting expulsion for documented misconduct violations.[55]

As in any other arena, the potential use of Non-Disclosure Agreements (NDAs) is controversial. Where harassment is thought to have occurred, but it sits below a criminal level of proof or where no formal investigation can establish the facts unequivocally or is indeed never carried out, often a settlement may be reached, putting a condition of silence on both parties. Again, this was the case with regard to Weinstein, where money to pay people off seems to have been no object. But an NDA can then lead to an alleged perpetrator being allowed to continuing working and continuing to offend. Sometimes the perpetrator is even enabled to move from one institution to another, without their new employer being aware of what has happened, which inevitably means no safeguarding will be put in place.

One victim of sexual harassment based in the UK, astrophysicist Emma Chapman, got clearance to speak out about her experience and has since actively been campaigning to have Non-Disclosure Agreements (NDAs) which silence victims banned in universities. She argues that pushing harassment under the carpet, as NDAs will do, damage the victims and give the perpetrators free rein to continue and create a toxic work environment for those who have partial knowledge of what is going on. There is still a huge amount of work to do to change the workplace culture so that anyone who has suspicions of what may be going on, or sees unsavoury behaviour, has the courage to speak up, as well as support from the senior leadership when they do so. It should not just be left to the victims, who already have too much burden to carry, to point out what is going wrong. As Sally Davies said to me in an interview about her time as a senior leader, including as the UK's Chief Medical Officer,

> *I think it's incumbent on those of us who are senior to call out [bullying] … I saw bad behaviour and I could see the people around either felt it was too difficult or it might impact on them. So I said, this behaviour won't happen in the future because I won't have it. And all the junior staff afterwards came and said thank you.*[56]

Field Trips

Harassment can occur anywhere, and what I have detailed so far is mainly relevant to lab-based science. However, not all science is done in the lab. Field trips are a crucial part of education and research in many fields, ranging from animal behaviour and ecology to geology and anthropology. Data-gathering in the field, which may be in very isolated places, provides another environment which can too easily turn toxic. This is true at undergraduate level, where such work may be a requisite part of the curriculum, as well as for postgraduate researchers and beyond.

An extensive study of field workers' experiences showed the magnitude of the problem.[57] Nearly three quarters (72.4%) of the respondents to an online survey reported that they had either directly observed or been told about the occurrence of other field site researchers and/or colleagues making inappropriate or sexual remarks at their field site. Men were more likely to report that comments never occurred, whereas women tended to report that such comments occurred frequently. 64% of women had personally experienced sexual harassment. Even more damning, over 20% of respondents reported that they had personally experienced sexual assault, described as physical sexual harassment, unwanted sexual contact, or sexual contact in which they could not or did not give consent, or felt it would be unsafe to fight back or not give consent. These figures are an appalling indictment of the environment in which many researchers find themselves and yet, if these people are to succeed in their field, such situations need to be faced and survived.

As the National Academies report spells out '*higher education is perceived, and in many cases accurately perceived, to tolerate sexually harassing behavior*'. A remote location for a field trip, with one or two powerful research leaders and a number of more junior researchers under their 'care', can provide a place where predatory behaviour is hard to check, and those at or near the top of that particular pecking order may believe, not without reason, that their behaviour will not be reported or stopped. Indeed, it would seem, from the report led by Clancy, that most individuals did not know what should be done or how to report inappropriate behaviour.[58] Fewer than half of those surveyed recalled ever seeing a code of conduct at any of the sites at which they had worked and less than a quarter believed they had ever worked at a site which provided a sexual harassment policy. Having policies in place which are well publicized seems the minimum that should be operating. However much the difficulties of using the codes in practice may deter would-be complainants, without such codes the situation is even worse.

The Net Effect

Each of the factors described in the previous sections, ranging from bias in letters of recommendation to outright harassment, leads to a steady drip-drip-drip from the pipe of progression. As the evidence builds about subtle and not-so-subtle forms of bias, perhaps there is some hope that it will become less 'unconscious', more visible in practice and therefore offering some chance of eradication. Better monitoring at every stage of the academic process will enable systemic biases to be identified and tackled.

Nevertheless, some of these factors may even get worse in mid-career, something that probably needs to receive more attention.

A woman who has had a good relationship with older colleagues may, as she moves from a junior to a more established position, stop being someone who can be patronized and suddenly become a threat. Mid-career women may find support lessens and outright hostility and confrontation increases. As a senior professor and sometime adviser to government put it to me by email:

> *A common theme I have heard from senior women colleagues in physics and engineering (still far too few exist) is that senior men see junior women as either fodder for harassment or intimate relationships or, if the men want to signal their 'allyship', they may appear to mentor and encourage junior women. Meanwhile they are coaching and sponsoring junior men. If those junior women manage to survive and become senior, those same senior men then see them as competition or a threat—both to themselves and to the junior men then have been coaching. This is when the most egregious and vicious discrimination and attacks come to the fore. These can take many forms—subtle/structural or more overt, but all act to limit the agency of senior women—as leaders in their own right but also in their ability to support junior women.*

For myself, I was surprised to find that denigration of my science and abilities actually got worse in mid-career, presumably because I was seen as that threat mentioned in the quote. It would seem that scientists will use any route they can to get ahead, seeing the lab as a zero-sum game. Not everyone encounters such issues. Other women I've asked say how much easier things become as they progress, that having an established track record makes them hard to challenge, or that they have developed a thicker skin and more resilience.

We've come a long way from the experience Jocelyn Bell Burnell was subjected to at Glasgow University where she studied for her degree in the 1960s, when 'tradition' meant that '*whenever a woman walked into a lecture hall all the guys in the room would slam their desks and whistle and catcall. Every time'.*[59]

But we haven't come as far as many might have expected. Nancy Hopkins feels too much is tolerated by too many:

> *Why did I allow myself to be treated this way for all those early years? I think that when you are young, you just have this driving energy, you love the science so much and you find excuses to put up with it all because you want to do the science .... I increasingly have trouble with established women who don't speak out.*[60]

As long as we fail to change the culture, the STEM career pipeline will continue to leak. Training women in STEM subjects who then find progression is slower than for their equally (or less) well qualified male colleagues is to no one's advantage, except possibly a mediocre man's. Minority status, put simply, is a handicap. The net effect is that many women do not get to follow their scientific dreams and have a wonderful career. And for those who do there may be punctuating moments of frustration and hostility to work through. So, what can be done?

CHAPTER 8

WHERE ARE WE GOING?

> I plead with all of my male colleagues that if you are asked to go on a panel at a meeting, or to give a talk, to ask what the gender balance is. If the organisers claim that they did ask some women, but they couldn't make it, then you say 'well, that's not for me'. If we make life better for female scientists, we make life better for male scientists too: a good working environment, child care, all those things are good for men and for women.[1]
>
> Anne Glover, former Chief Scientific Adviser to the President of the European Commission

> Don't leave this wonderful, fun work just for the men.[2]
>
> Frances Arnold, Nobel Prize winner in Chemistry

All professional sectors seem to have issues with gender equality, be it accountancy or advertising, law or medicine. However, the STEM subjects, particularly the physical sciences, engineering and technology, seem to have particular issues because young women are so often deterred from ever putting their foot on the STEM ladder at all. Even if they succeed and go on to gain their independence, the previous chapters show that the playing field can still be far from level and working environments may remain unfriendly, even hostile and toxic. How can all this be changed?

We should ask what equity would look like. It needn't, I believe, mean equal numbers of men and women students in every discipline, but the wild variations in many subjects seen in the Western world seem unlikely to be down to innate differences, given that other cultures do not exhibit the same differences. Indeed, there is something of an inverse relationship between the number of women choosing science and the wealth of a country. Table 1 shows data from selected national academies of science, listing those with women representing 20% or more of members in 2020.[3] The UK and the USA national academies fall just outside this list, both with 19%, although the US National Academy of Medicine does appear.

It will be seen that many of those listed are in Latin America and the Caribbean. Despite many of these being small academies, it is striking how high they appear on the list. These differences will, however, be influenced by the longer-established academies (such as many of the European ones and those of the USA) having many more older scientists in their fellowships, elected at a time when there were far fewer women in the scientific workforce: younger academies will only have members elected in recent years, when numbers of women have moved a little closer to parity. Of the Scandinavian countries, which might be expected to have high proportions of women given their reputation for gender equity, only Norway is on the list, and it only scrapes in at the bottom.

Looking at the total researcher population in different parts of the world shows a similar story, as shown in Figure 10. Here it is clear that both Central Asia and Latin America and the Caribbean have far higher proportions of women in the researcher population than the USA or Western Europe.

Equity means, to me, the opportunity for everybody to be free to make the career choices that are best for them, not what other people's expectations force upon them. It is also important to recognize that a woman who drops out at child-bearing age may be

Table 1 Percentage of women members of those national academies which have at least 20% female members [1]

Academy	Country	2020 survey		
		Total members	Women members	% Women
South African Young Academy of Science	South Africa	49	28	57%
Young Academy Finland	Finland	74	41	55%
National Academy of Young Scientists	Pakistan	200	110	55%
Young Academy of Belgium	Belgium	49	25	51%
Hungarian Young Academy	Hungary	24	12	50%
Association of Latvian Young Scientists	Latvia	154	74	48%
Global Young Academy	Germany	200	89	45%
Young Academy of Europe	Germany	135	60	44%
Die Junge Akademie	Germany	50	22	44%
Polish Young Academy	Poland	32	14	44%
Academy of Sciences Cuba	Cuba	375	122	33%
Koninklijke Academie voor Geneeskunde van Belgie	Belgium	84	26	31%
Academia de Ciencias Fisicas, Matemáticas y Naturales de Venezuela	Venezuela	42	13	31%
National Academy of Sciences Honduras	Honduras	35	10	29%
Nicaraguan Academy of Sciences	Nicaragua	36	10	28%
National Academy of Medicine	USA	2242	631	28%
Science Council of Japan	Japan	2210	609	28%
Caribbean Academy of Sciences	Trinidad and Tobago	150	40	27%

Academy of Science South Africa	South Africa	573	155	27%
Koninklijke Academie voor Nederlandse Taal en Letteren	Belgium	31	8	26%
Royal Society of Canada	Canada	2273	558	25%
Academia Mexicana de Ciencias	Mexico	2832	704	25%
Lebanese Academy of Sciences	Lebanon	29	7	24%
Koninklijke Vlaamse Academie van België voor Wetenschappen en Kunsten	Belgium	295	71	24%
Cameroon Academy of Young Scientists	Cameroon	40	9	23%
Academia de Ciencias Medicas, Fisicas y Naturales de Guatemala	Guatemala	87	20	23%
Slovak Academy of Sciences	Slovakia	48	11	23%
Academy of Scientific Research and Technology	Egypt	600	125	21%
National Academy of Sciences of Sri Lanka	Sri Lanka	146	30	21%
Academia Nacional de Ciencias	Peru	116	23	20%
Latvian Academy of Sciences	Latvia	416	85	20%
Norwegian Academy of Science and Letters	Norway	931	188	20%

[1] 'Gender Equality in Science: Inclusion and Participation of Women in Global Science Organizations. Results of two global surveys.' September 2021, Interacademies Partnership, https://www.interacademies.org/sites/default/files/2021-10/Gender%20Equality%20in%20Science.pdf.

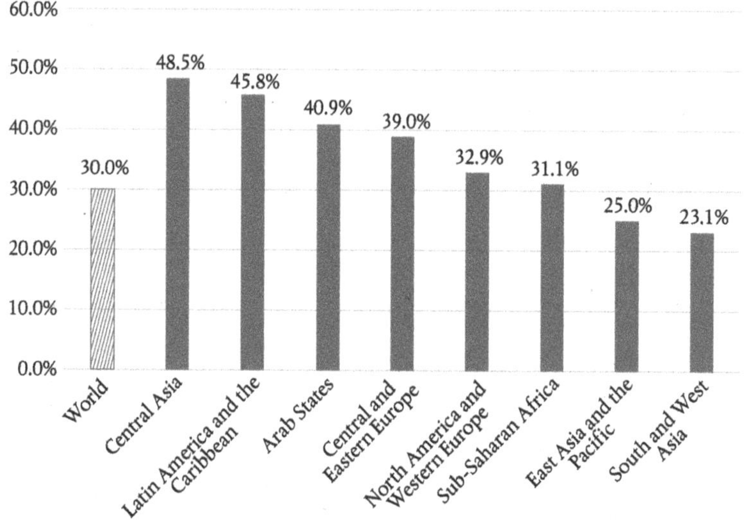

Figure 10 Data from UNESCO Fact Sheet 60, 'Women in Science'.
Magdelena Szmigiera, 'Share of female research & development (R&D) researchers in 2017, by world region', *Statista*, 21 January 2022.

doing it not from choice, but from a subtle combination of factors making her feel it's no longer worth the effort to keep battling on. Families may be used as the 'excuse' to explain why women don't progress at the same rate, but the argument that they no longer have the wish to continue should be recognized as a false if convenient myth in many cases.

Is progress being made? Yes, but the odds are still stacked against the typical woman. In the 21st century we need more women entering the STEM subjects, contributing to the solutions we need to the problems of this changing world and sparking the economy through innovation and disruptive technologies. To do this we need, across the world, to enable more women to enter and flourish across all the STEM professions, most notably the physical sciences and engineering, by eradicating the barriers society has imposed on them and discarding the cultural expectations of our gendered societies. There are many relatively minor changes which could make a substantial difference at the societal level. We see increasing examples of role models to inspire future generations—women such as Fabiola Gianotti and Donna Strickland, both of whom I've mentioned earlier—and we need to see such women appearing in the media on a regular basis, not just when a Nobel Prize is awarded. There are various initiatives within the UK that seek to achieve this. Notable in this space are two different examples from the BBC. The journalist Ros Atkins started up a project for his news programme which he called the 50:50 project, aiming to achieve 50% of the experts and journalists used in the programme being women. This project has now spread much further across a wide range of programmes.[4] That initiative spans many areas of expertise. More specifically in the science domain is physicist's Jim Al-Khalili's radio programme, *The Life Scientific*. It may not be publicly written down anywhere, but his approach, which he told me at the time he signed me up for an interview, is to ensure an equal number of male and female scientists are interviewed about their life and science. All the podcasts are freely available to anyone who has access to the BBC website,[5] so that there is a lasting pool of individual stories of a diverse range of female scientists to dip into.

School and college textbooks need to contain the discoveries, stories and images of women. Florence Nightingale was a statistician at least as much as she was a nurse, but that fact doesn't make it into popular culture or the textbooks children see, although almost every English child will be familiar with the story of the 'Lady and the Lamp'. A recent study by the American Chemical

Society examined 10 standard US Chemistry college-level textbooks to look at the gender representation of scientists.[6] On average, females were found to constitute 30% of images and 3% of the named STEM professionals in the index, and whereas a male name cropped up on average in every four pages of text, a female name appeared only once in every 250 pages of text. No one may have set out to exclude women, but clearly they didn't try that hard to include even present-day women to counter prevailing stereotypes. Few female role models there for students to relate to.

TV and streaming channel programmes and films need more women in technical roles, not just as the love interest. A 2018 study from the Geena Davis Institute on Gender in Media looked at the impact of X-Files' protagonist Dana Scully on girls and women who watched the programme: half said their interest in STEM increased because of her character, and 63% of the women in STEM they interviewed regarded her as their role model.[7] A follow-on study from the same institute in 2021 found that over 80% of the participants believed it was important to see women STEM characters in film and television. Of the girls and women planning on pursuing STEM careers, a substantial majority claimed that STEM characters from such shows inspired them to do so.[8] However, the report also noted that nearly 90% of actors seen in STEM roles in films around the world were in practice male. Their simple recommendation, unsurprisingly, is that the media needs to move away from almost invariably portraying scientists as white men.

TV and films need to showcase men and women scientists as typical, not to mention sane, human beings; instead, there are far too many examples of 'geeks', using that word in the negative sense of dysfunctional or mad scientists. I've long thought more variation in the kinds of jobs and professions soap opera stars follow would help to normalize some of these roles, rather than any scientists who are introduced into a script portrayed as very not 'normal' (or else simply invisible). *Hidden Figures,*[9] the book by Margot Lee Shetterly and the subsequent film version, did more than highlight the race dimension, or even intersectionality; it showed that key individuals, such as Katherine Johnson and Mary Jackson, had ordinary backgrounds without STEM exemplars in their families, and had regular families of their own, but they still could do and did extraordinary things for NASA.

The other side of the coin relates to practical steps that could be taken to reduce the hurdles women need to overcome if they are to progress in STEM. Many of these are not STEM-specific and would make a difference in many sectors. Actions that could be taken would be to make sure that childcare costs were affordable; this is particularly important for early career researchers. Also for parents, wider availability of on-site crèches would be a great benefit, especially for mothers during the early months of feeding. Easy access to flexible working and 'emergency' days of leave to be taken when family emergencies arise, would give confidence to carers. In the UK, shared parental leave is available in many situations but, as a 2020 Chartered Institute of Personnel and Development report made clear, around three quarters of men believe there is a still a stigma attached to a man taking extended periods of such leave, and nearly all men agree that there would need to be a real workplace culture transformation for this to be normalized.[10] Additionally, particularly if the man is the higher paid in a couple, there may be significant financial implications if the man takes the extended leave to which he is in principle entitled. Consequently, the take up remains stubbornly low, defeating the laudable aims of the policy. If parenting (not just mothering) is to be taken seriously, it is imperative that men share the responsibilities from the outset.

Sometimes I am asked, why does it matter if girls 'don't want to do science?' Implicit in that question is the belief that girls really *don't* want to do science, rather than that they imbibe the message that they shouldn't. The UK's Parliamentary Science and Technology Select Committee, carrying out an inquiry in 2022 into diversity and inclusion in STEM, asked a well-known secondary school head teacher, Katharine Birbalsingh, why girls were turning away from Physics.[11] Her response, '*physics isn't something that girls tend to fancy. They don't want to do it, they don't like it,*' caused something of a furore in the media, since in those words she ignored all the evidence I've gathered together in this book to demonstrate why culture imposes its own way on girls' choices. Indeed, she made it clear she was unaware of most of this evidence when she said: '*The research generally … just says that's a natural thing … I don't think there's anything external.*' She did not even appreciate that her school's statistics were worse than the average in the percentage of girls taking physics at A level.

Her answers to the enquiry, though, highlight just how important it is to get the message out that the reason physics lacks girls is not that it is 'natural', or that girls are put off by hard maths. There are presumably many teachers in schools around the world who, without giving the matter much thought, inherently believe such cultural stereotypes and act upon them unconsciously. This book aims to shift the dial a little, but there are many voices out there stressing what the evidence genuinely is if (head)teachers choose to look. I was very pleased to be given the opportunity to give my own evidence to a later session of this same Parliamentary enquiry, stressing all the cultural barriers I've identified in the previous pages.[12] Explicit actions to deter girls at school are probably less common now than decades ago, but as long as teachers, no doubt with the best of intentions, give out these sorts of messages, generations of girls will continue to feel discouraged and that STEM is not for them.

In this book I've used quotes from a number of Nobel Prize winners and, as the number of women winning these prizes increases, it has to be hoped that their names will find their way into textbooks as a matter of course. Nobels only represent the tip of the scientific iceberg, of course. They are divisive in many ways, not least when it comes to choosing who should win from a team effort, but they are firmly ensconced in the public's eye and so their winners are more memorable than many other equally outstanding, inventive scientists who could make strong role models too. Covid-19 brought many women into the public eye. Women such as Özlem Türeci (one of the co-founders of BioNTech, the company behind the Pfizer vaccine) and Sarah Gilbert, who led the team developing the Astra Zeneca vaccine.

It is to be hoped these women will feature in future textbooks, along with many others. In the UK it has been noticeable how many of those writing and speaking about the pandemic from the viewpoint of their various specialities have been women. Susan Michie, daughter of Anne Maclaren, who I've mentioned before, and who is a psychologist, has been outspoken within the UK press about the need for caution in emerging from lockdowns. Devi Sridhar, a public health expert at Edinburgh University, Charlotte Summers a critical care specialist from Cambridge, and Oxford-based palliative care medic Rachel Clarke have all been vociferous in the media about government policy and the impact on the ability to provide appropriate care in hospitals as the numbers of critically ill individuals mushroomed. Sridhar, a Professor of Global Public Health, acted as an advisor to the Scottish government as they developed their strategies and wrote extensively about the pandemic in the media, following up with a book about the response around the world.[13] It is true these are all women whose expertise sits close to the medical professions rather than the physical sciences, but in terms of a very visible presence in scientifically driven subjects, I hope they make a difference in public and parental perceptions about the desirability and accessibility of science for girls and women.

We cannot rewrite history to insert more women into our scientific heritage, but we can (and should) celebrate those who did make notable contributions in spite of the odds. Women can so easily be written out or overlooked, as has clearly happened in the case of female composers, who are finally being 'discovered': think Ethyl Smythe, Florence Price or Amy Beach around the turn of the 20th century, or Barbara Strozzi or Françoise-Charlotte Ménétou from the seventeenth. Even more importantly, we should remember the women of the recent past whose works are notable but too often, via the Matilda Effect, attributed to males, such as Lise Meitner. These women should be brought out of the shadows in our education system, to inspire future generations. Furthermore, it should not just be those women who go on to successful academic careers who are highlighted. Although this book has paid most attention to such a trajectory, we need scientifically trained women in a far broader range of roles and professions. We need those who bring their scientific way of thinking into a wide range of different spheres, including as politicians and as concerned citizens.

The big challenges we face, such as creating adaptations to allow us to live with climate change or technologies to facilitate moving to a carbon neutral economy, require all the creative scientific thinking input that can be mustered into the policies governments around the world may mandate. The importance of scientific input into policy decisions during the Covid-19 pandemic has been manifest: questions such as should face coverings be worn and where, should children as young as twelve be vaccinated or how long should quarantine be imposed? These are key questions which different countries have addressed differently. When balancing scientific advice with wider issues including the economy and safeguarding jobs, it is the politicians who have to make the decisions, but they need to ensure they get the best scientific advice they can. Women will bring their own perspectives to all these problems, perhaps because they are the primary carer, or (in some parts of the world), the one who is more used to having to fetch water or cook on wood fires. Existing societal and cultural pressures mean what matters to them may be different from a male perspective, and therefore they may attempt to solve problems that men may overlook, or solve them via different approaches. This is a reflection of the same issues I identified in Chapter 2 when discussing why diversity matters. Better solutions will be found if a diversity of individuals is empowered to contribute.

When I chaired the Scientific Advisory Council for the Department of Culture, Media and Sport (DCMS) some years back, it appeared that there was only one scientist amongst all their teams of civil servants, but plenty of economists. I felt that the absence of a strong voice aware of scientific ways of thinking and developing ideas created a huge gap.[14] For instance, how can one quantify how much sport really makes a difference to a growing child's health? Or what are the developments in digital archiving relevant to museum collections, including for items 'born digital', and should these systems be standardized? Those are just a couple of specific examples. Joining the Civil Service in the UK, or equivalents elsewhere, is an important way in which scientists can make a real policy difference. The same applies to working in local government, where issues such as local flood protection or soil remediation may become political hot topics and need scientific input. Additionally, there are opportunities at think tanks (national and international) where policy papers are written, which help to define the policy agenda or provide analysis to feed into other institution's developing ideas, such as the professional bodies and learned societies. Here too, women's voices need to be heard to provide a well-rounded view with diverse perspectives.

In a different sphere, careers based on talking and writing about science, enthusing about science, passing on the love of science to the next generation in schools, through festivals and outreach, should be promoted. Some of those who do this in print and on screen are academics. Examples of these might include the Harvard professor and theoretical physicist Lisa Randall and the writer and TV presenter Alice Roberts, trained as a doctor, now part-anthropologist, part-biologist and a Professor in the Public Understanding of Science at the University of Birmingham. However, every scientist working in a museum or school, running a science festival or penning a blog, is making their own contribution in a meaningful way, using their knowledge to inspire others. Women as role models matter. These voices, expressed in appropriately non-technical ways, are vital in ensuring the public are exposed to relevant scientific ideas in accessible ways.

As the Royal Society wrote in a 2014 report on the future of education '*Scientific and mathematical understanding is fundamental to fostering an inclusive and effective democracy*'.[15] Once again, the pandemic has brought this to the fore, where those segments of society around the world who aren't convinced by the data regarding infections or the safety of vaccines, have put themselves and others at risk of serious ill health and death. Of course, teachers are a crucial part of the scientific ecosystem, vital for nurturing and enthusing talent in the early years. Having women teaching the STEM subjects is yet another way of providing role models, and some evidence exists to demonstrate a very direct interplay of the number of female STEM teachers in schools to the numbers of female students who go on to take STEM at university level, being most pronounced for those with the highest maths skills.[16]

Un-Gendering of Science

Nineteenth century scientists may have believed that letting women study science—indeed practically any subject at a serious level—might lead to their wombs withering up and their health failing, but we have at least moved on from that position. Nevertheless, despite all the accruing evidence that there is no inherent distinction between a male and female brain, somehow our society's belief that there are boys' subjects and girls' subjects persists, hurting boys too. However, for centuries, the education of girls has been systematically restricted, and discrimination against girls' moving into counter-stereotypical arenas both formally and informally enforced. Education systems should be designed to bring out the best in an individual, not attempt to fit them into some mental and gendered iron cage.

The question then is not whether girls *can* do science, but how we ensure, through cultural and societal change around the world, that those young women who want to do science (particularly the physical sciences, engineering, computing and mathematics, where the numbers are lowest) feel free to make those choices and follow their interests. Interventions in later adolescent and college years have been tried and ultimately, they appear to make limited difference. My belief is that we need to start changing attitudes directed towards the youngest children. There is no need to ban Barbie dolls, or pinkification, or Disney princess dresses as gifts for the young girl to make this difference. Instead, we need to ensure that alongside these sorts of toys we provide other types that encourage creativity and thinking in three-dimensions: for instance, these could include design and construction sets, such as Meccano, chemistry sets, and Rubik's cube. Such toys will fuel imagination as well as spatial awareness. We need to add into the mix of upbringing and messaging a willingness to embrace risk, to try out and fail at everything from climbing trees to winning arguments, an acceptance that nice girls can and should be empowered, confident and daring without losing respect and friendships. It goes without saying I believe the range of toys offered to boys and the types of behaviour regarded as appropriate for them, should also encompass a much wider range. Gendering attitudes towards boys are every bit as pernicious as towards girls, equally limiting their life choices subsequently.

Having offered both boys and girls, from their earliest days, a wide range of toys, we also need to consider their reading and viewing matter, and this is where the need I've mentioned for more images of women in STEM subjects in textbooks and broadening the roles of scientists in drama and films comes in. The lack of images of women in scientific roles in school and college textbooks could easily be rectified. There are biographies of women scientists now appearing on the market for all ages, but my suspicion is these are given (or read) to girls, and not to their brothers. It is important that boys as well as girls realize that scientists are people: not men, not women, but both, either or non-binary. As long as boys continue to believe that 'girls can't do science', peer pressure will play its part in deterring the less confident girls from following their dreams.

Career Trajectories and Families

The leaky pipeline metaphor may not be the best way of representing what happens to women, but it certainly highlights the fact that far larger numbers of women leave the STEM workforce than men, often without even taking up a first job in a relevant area. I am firmly of the view that if they leave because they've decided that it no longer aligns with their aspirations, well and good. If they leave because they find the environment unconducive to women specifically, and to families in general, then there remains a problem. Given that the rates at which women leave different disciplines vary quite substantially, the disciplinary differences in environment can clearly be subtle and hard to unravel.

The challenge, as in so many professions and jobs, is how to make the career ladder manageable for those who want a family, men and women. For wider society, the problem is that those who leave may be every bit as smart as those who stay. Their loss is a loss to us all. It is worth asking whether the precarity of early career positions so often encountered and the requirement that an individual does not stay put in a single lab really are the best sifts of talent. This is a conversation that may be starting in some places but hasn't gone nearly far enough. The model for academia was created in an age when science researchers were male-by-default, along with the expectation of such a man having a wife who stayed at home. No other career, to my knowledge, has these particular obstacles. Postdoc precarity is driven by the way projects are funded. Many funders award grants for only two to three years (some award five-year grants, which in principle could support individual postdocs for longer, although that may not happen), and institutions then offer contracts for that duration. Taking a risk that a follow-on grant will be awarded to enable the postdoc to stay on is something few institutions are willing to do. If they did, if postdocs were promised a job for life even without progressing to professorial ranks, would that solve the problem? I fear not, as what would happen to all the PhD students who came thereafter?

This problem is one of supply and demand. A 2014 analysis of the situation in the USA pointed out that '*the system in many places is saturated, far beyond capacity to absorb new PhDs in academia at the rates that they are being produced*'.[17] Countries treat training of PhDs as if it were independent of the subsequent pyramid of tenured ranks of the professoriat. If precarity and the consequent problems this causes for women at this career stage could be removed by a complete rethink of academic structures, then perhaps the leaky pipeline would leak less. At the moment, it seems that far too many of those who start PhDs are unaware of the uncertainty of their future careers, and don't look beyond academia despite the attractiveness of all the skills they gain to many other employers. It is true that the high profile major industrial laboratories with a 'blue sky' remit,[18] exemplified in the past by Bell Labs, and the large laboratories at companies such as IBM (in the USA) and ICI (in the UK), have often either been totally closed down or reduced in scope. Some government laboratories have also been shrunk or, in the UK, privatized. Nevertheless, there is still a great need for scientists, technologists and engineers outside academia, to join many companies large and small, long-established or a start-up, as well as in government laboratories. The opportunities have not gone away, although perhaps they are in less monolithic prominent organizations.

Environment

Over and above the specifics of how a career unfolds, the whole question of the working environment needs to be addressed. Unconscious bias training, as I noted earlier, has become very popular as an apparently easy fix, but is proving inadequate as a means of removing deep-rooted beliefs and may even backfire.[19] In order to create a more inclusive and welcoming environment for everyone, much more studied and long-term approaches are needed.

In my view, we need to move away from working environments that promote the cult of the individual big-shot team leader whose behaviour is considered untouchable. Although proven cases of sexual predation in academia remain relatively uncommon (as opposed to rumour, anecdote and suspicion), the fact that the US National Academy of Science has gone so far as to expel a member because of his (proven) sexual harassment is a welcome step, making a public statement that such behaviour will not be tolerated, however impressive the individual's research record. That the funder, the Wellcome Trust, has introduced a policy of the withdrawal of funding from those who have been found guilty of bullying is another positive sign. For too long, bullying and harassment have occurred relatively openly but without sanction. Science may be inherently competitive, as different groups chase a specific target. Nevertheless, it doesn't have to be conducted in a toxic environment.

In this context, I would highlight two specific actions anyone can take: amplification (discussed earlier) and allyship. Every committee around the world, in academia or not, could do with more pushing back on bad behaviour from other committee members to counter the 'Miss Triggs' effect of the Punch cartoon, shown in Chapter 7. If someone hears another member's ideas being appropriated by someone else, they can step in. A line such as '*I'm so glad you agree with Mary* …' can help to remind everyone else whose idea it originally was. Even without another person attempting to appropriate the idea, amplification can still come into play simply by reinforcing the message of who made the key point. The moral is, don't let a good idea be stolen, don't let the Matilda Effect come into play, but do everything possible to ensure due acknowledgement is given to the originator, however lowly their power status. Likewise, the other tendency seen only too often at committees, debates, and other potentially confrontational situations, is for men to talk over another committee member, typically, but not necessarily, a woman. A good chair (male or female) will step in to stop this, just as they will facilitate the quieter less confident members to speak up. In that way a chair can crucially act as an ally, but too often does not.

Allies in the context of any situation involving a victim, are always helpful. I would go further and say that, if a group member observes inappropriate behaviour of any sort and does nothing, they are complicit in permitting that behaviour to continue. Power differentials may make it hard for a junior member to do much. Nevertheless, at the very least privately acknowledging to a victim that something inappropriate has indeed happened, that it wasn't simply in their imagination, and checking if the affected person is doing OK, is a part that anyone can play, however junior.

The idea of 'men as allies' could include the use of amplification, but as research from the neighbouring college to my own in Cambridge, Murray Edwards,[20] has shown, there are many ways in which men can support women. Their research, summed up in two booklets under the *Women Collaborating with Men* banner (*Inclusive Networking and Sponsorship*[21] and *Everyday Workplace Inclusion*),[22] is not directed at academia, let alone scientific academic departments, but the concepts apply equally well here. When an individual is part of a disadvantaged minority, of whatever kind, being an 'ally', supporting them, sticking up for them, or drawing opportunities to their attention, will always be of assistance. In most science departments, where women are likely to be in a minority at any stage past the first degree stage, men acting as allies will help to facilitate the progression of women. The concept of allies is relatively new, but I believe crucially important. Alongside being an ally could be a willingness to act as a mentor or sponsor. Men need consciously to stand up for this role, and not (unconsciously or otherwise) assume men should sponsor men, and perhaps women should sponsor women. I do not believe that you have to be the same sex to act in that role in general.

Increasingly, organizations are turning to so-called Behavioural Competency Frameworks (BCFs), to spell out what sorts of behaviour are expected of colleagues during processes such as recruitment and progression, as well as in daily life. These do not yet seem to have been significantly adopted in universities, although some public sector organizations are operating with such frameworks publicly available. A recent communication from three senior women scientists has proposed these should be widely used in STEM environments, based on the experience of one of them working in a Wellcome-funded institute which operates with one.[23] The point of a BCF is that it specifically sets out how staff members are expected to behave when employed by an institution. Publicly setting out the norms serves to help people address their own conscious and unconscious behaviours and biases. Such frameworks are akin to codes of conduct many scientific organizations utilize to govern behaviour at conferences. One advantage of such expectations being explicitly set out is that it also enables routes for complaints to be detailed, so that no one is in any doubt about who to talk to and how complaints will be handled. As I indicated in the case of field work, the existence of such frameworks or equivalent codes of conduct do seem a minimal expectation for any organization to put in place, yet often are not.

Am I Good Enough?

The exclusions I have talked about so far are provoked by circumstances, or by other people. But there are also self-exclusions, due to feeling that one's credentials don't match up to the opportunity

on offer or that one wouldn't fit in given how organizations are organized. There is increasing literature regarding what might make a job opening inclusively attractive. For instance, the use of gender-neutral job descriptors is essential and by now widely adapted. The less obvious deterrent effect of gendered words stereotypically associated with personality traits is also being increasingly addressed. Words such as ambitious and assertive appear in advertisements but, as with their appearance in letters of reference, would stereotypically be associated with males. One detailed analysis of job advertisements demonstrated that advertisements for male-dominated occupations (which would include many STEM-related occupations) contained more stereotypically masculine words than job advertisements for female-dominated occupations.[24]

A recent analysis using machine learning approaches nevertheless indicates that removing gender-explicit words from job advertisements is insufficient to solve the problem because of this association of traits with gender.[25] But it is often hard to find more gender-neutral equivalents to words such as ambition/ambitiousness. In this paper the authors discuss an algorithm which not only evaluates gender bias in the input text, but also makes suggestions about alternative wording which is less biased while maintaining the sense. In principle this should improve the appeal of advertisements to a wider range of applicants and the approach offers a quasi-solution, but only as long as HR departments are willing to choose to use algorithms when constructing job applications. At the very least, I hope this aspect of recruitment receives a lot more attention in the future, beyond mere 'boiler plate' text encouraging applicants from a diverse range of backgrounds to apply.

I believe consciousness-raising regarding this sort of non-deliberate bias is needed. I was largely responsible for UKRI re-issuing an advertisement for Board members due to the singularly off-putting language used. It is to their credit that, when the deficiencies were pointed out, the agency was willing to start again. As I wrote about this episode:

> *Some folk may actively drive, speak authoritatively and head off in pursuit of their mission—as apparently explicitly required by the person specification that appeared—but it isn't necessarily a vocabulary everyone is likely to use. Indeed, many people, men and women, may not be comfortable with thinking of themselves in those robust phrases. I know of some very senior women from the top echelons of Russell Group universities who told me they looked at the advertisement and decided it 'wasn't for them' because of the language in which the advertisement was couched and its whole tenor. That, to my mind, meant that it had failed on a crucial front.*[26]

This problem, like many others, is not specific to the STEM disciplines but, when numbers of women moving up the career ladder are anyhow low, this sort of hurdle will simply restrict the progress of women, as well as more modest men, to the top, without it being by anyone's conscious design. Looking hard at the wording of a job advertisement ought to be a straightforward task that any organization could undertake in the natural course of their work, but it is one which has the potential to widen the field of applicants substantially. Those producing glossy recruitment further particulars, commonly used for high level jobs, have become increasingly conscious about including images showing diverse members of their community. This is something that has changed over the past decade or so; it ought to be possible to get the wording of the actual advertisement to be equally inclusive.

Where Next?

Over my career I have seen enormous changes in the way the question of women in science is addressed. When I was setting out it wasn't much talked about at all, as I recall. There were, in the UK at least, no formal barriers. There were of course informal ones, such as caps on the number of women Cambridge University would admit, but if I wanted to be so weird as to study physics, it was open to me to try to do so. I might have ended up—as Jocelyn Bell Burnell did a few years before me at Glasgow University—being catcalled as I walked into a lecture theatre, but luckily for me I wasn't. I might have ended up, as she did during her PhD in Cambridge, being effectively written off as a serious researcher who had just made the stunning discovery of pulsars, when she got engaged[27]; in fact, my engagement and marriage during my PhD did not seem to impact on the attitudes of those around me. No one told me I should give up aiming for a postdoctoral position just because my personal circumstances had changed. I was, as I've said throughout this book, lucky on so many fronts. But the world I was moving through was changing.

By the time I had my first child, I was able to get some weeks paid maternity leave from my employer, but there were practically no prior examples across the University, and certainly none in my department. How I organized my life thereafter, how I coped with childcare, was such an unfamiliar situation no one asked me any questions. My husband and I made it up as we went along, because there was essentially no one to discuss it with. I believe that was helpful, since my choices weren't scrutinized in the way women's choices are these days. That we worked out ways to share the caring between us, and that in the end it was my partner who gave up his formal career rather than me, was even more unusual. His decision to become a house husband while continuing with his research was a choice only possible because of the nature of his research and the comparative cheapness, at the time (certainly no longer true), of houses in Cambridge, allowing us to live close to the city centre. Nevertheless, it did mean the end of his salaried career, a hard choice for anyone to make.

Starting a family, therefore, did not completely upend my career, as it does for so many women. Things may have looked like they were improving on this front, with more organizations providing childcare on site, and better financial support available in some places. However, the Covid-19 pandemic has thrown everything up in the air again. How serious the consequences for some women will be due to the way childcare has impacted on their ability to work during the height of lockdown and how this may echo down the years ahead is unclear. This was identified as early as 2021 as problematic by a report from the National Academies of Science, Engineering and Medicine.[28] It is too early to be able to identify all the consequences of this global disruption, but it is hard to see them as likely to be positive for women in science.

Issues such as childcare that had barely surfaced when I had to face them are commonly discussed now. But such open discussion undoubtedly has pros and cons. On the plus side, there are places to go for advice, other women who've tried this or that route and websites to assist. No one need think they are doing it alone or are odd for wanting to combine science and family. However, the downside is that it isn't possible to be as flexible as I was, when flexible or part-time work wasn't formally available. What I did in my days of flexible working really fell beneath the radar: if I did what was needed (particularly in the very visible form of teaching) and did it well, no one thought to ask me how I was spending the rest of my time. Health and safety also put more stringent demands on the individual today, usually entirely reasonably. While I managed to carry out an interview for a postdoc with a screaming baby on my shoulder back in the 1980s, others can come up against policies designed to prevent children from entering a laboratory. Even before my day, the late US physicist Millie Dresselhaus, informally referred to as the Queen of Carbon, faced this when she gave birth to her fourth child, as reported by her fellow physicist, Gene Stanley[29]:

> *When she had her fourth kid, she brought him to work the day after he was born. She was there around noon or 1 o'clock with the baby in tow. But because Lincoln Lab was a government lab, you either had to have clearance or have a badge. They wouldn't let the kid in. She was furious! I didn't see her angry that often, but I saw her angry that day.*

Some institutions can be anything but flexible; safety concerns have only got more stringent since then.

A decision has to be publicly made now about any part-time strategy. Presenteeism is likely to be regarded as important by some colleagues and the impression I get is that most people who officially work part-time usually end up working far more than their contracted hours; they simply don't get paid for the additional chunk of time. Nevertheless, the idea that meetings should be held in so-called core hours (broadly coinciding with the school day) is now commonplace in the UK, and certainly happens at least in some other countries to a greater or lesser extent. That key meetings should not be arranged during school half-terms is no longer an alien concept, whereas for me, it was always a matter of embarrassment trying to find acceptable excuses that didn't mention children when I found it hard to turn up at such times. So, these last two topics definitely represent progress. Other changes in workplace practice are still working through as a result of hybrid or 'agile' working post-pandemic, but these too may facilitate improved work-life balance.

When I set out, if a woman was harassed at a conference, that seemed just to be taken as normal. There was no one (or route) to report such harassment to formally (although, as I've indicated, I did manage to get my own unpleasant episode dealt with informally); it just had to be accepted. Shocking stories still abound, but there is at least some possibility of reporting them, with the introduction by some (if not by any means all) organizations of formal codes of conduct and protocols for how to deal with transgressions. How much effect this is having I'm not clear, because certainly drunken lecherous behaviour in the bar does not seem to have vanished, as far as I can tell. But at least such perpetrators know their behaviour may be challenged, possibly even publicly.

The token woman on a committee—that happened to me often enough—is no longer so common, because most organizations recognize having a single woman cannot change anything except her workload. However, some organizations do a better job than others of ensuring women are on the committees that make the big decisions, as opposed to the ones that do the chores. Women's role in leadership has progressed hugely, but has not achieved parity. In part this sits with those who select committee membership, in part the fact that women may either be hesitant to put themselves forward and/or are not tapped on the shoulder in the way some golden boys may be. Additionally, women, since they are so often in a minority, may not want to take on some of the roles that are offered to them, because of the impossibility of workload this may impose. Nevertheless, this is where all the words I've written about sponsorship and mentorship are so important if women are not going to continue to be held back from some roles, as well as access to some funding. I believe this is an issue still to be ironed out.

Many of the matters I've raised here will need to change at the institutional level. For the individual who wants to make a change, the checklist I wrote some years ago (see Table 2) covers actions I thought everyone, whether familiar with STEM careers or not, could flick through and find something they could do to support women in science.[30] I called it 'Just One Action for Women in Science' (#Just1ActionFWIS). For parents or friends, for senior leaders or for bosses, for policy makers or teachers, I felt there were easy actions everyone could take. I hope each reader will contemplate what *they* could do to ensure we don't lose so much talent.

When Can We Stop Talking about These Issues?

I believe the answer to this question is when mediocre men and women are equally likely to succeed—in other words, not soon. The *Financial Times* correspondent Pilita Clark put it neatly, in words applicable to any sphere, when she said: '*Women must demand the right to be as useless as men*'.[31] We must move on from a time when a woman has to have far more papers or patents to her name to progress than her male contemporary, or has to have more funding, more PhD students—just more of everything—to succeed. Women, in science as elsewhere, should not have to battle with the 'likeable versus competent' conundrum that hampered Hillary Clinton's presidential aspirations, and that has impacted on so many women in so many spheres.

There may be no explicit glass ceilings in most developed countries, no firm limits or explicit extra hurdles to overcome, but that doesn't mean we've reached a level playing field. Until our society recognizes that attitudes to a person's gender are shaped by social and societal expectations and that stereotypes are inappropriate and outdated fictions, we will not have equity. The conversation is happening but, as is inevitable, there is pushback. Those

Table 2 Athene Donald, 'Just one action for women in science', *The Guardian*, 19 June 2015, https://www.theguardian.com/science/occams-corner/2015/jun/19/just-one-action-for-women-in-science.

- Call out bad behaviour whenever and wherever you see it—in committees or in the street. Don't leave women to be victimized;
- Encourage women to dare, to take risks;
- Act as a sponsor or mentor;
- Don't let team members get away with demeaning behaviour, objectifying women or acting to exclude anyone;
- Seek out and remove microinequities wherever you spot them;
- Refuse to serve on single sex panels or at conferences without an appropriate level of female invited speakers;
- Consider the imagery in your department and ensure it represents a diverse group of individuals;
- Consider the daily working environment to see if anything inappropriate is lurking. If so, do something about it.
- Demand/require mandatory unconscious bias training, in particular for appointment and promotion panels;
- Call out teachers who tell girls they can't/shouldn't do maths, physics, etc.;
- Don't let the bold (male or female) monopolize the conversation in the classroom or the apparatus in the laboratory, at the expense of the timid (female or male);

- Ask schools about their progression rates for girls into the traditionally male subjects at A level (or indeed, the traditionally female subjects for boys);
- Nominate women for prizes, fellowships, etc.;
- Tap women on the shoulder to encourage them to apply for opportunities they otherwise would be unaware of or feel they were not qualified for;
- Move the dialogue on from part-time working equates to 'isn't serious' to part-time working means balancing different demands;
- Recognize the importance of family (and even love) for men and women;
- Be prepared to be a visible role model;
- Gather evidence, data and anecdote, to provide ammunition for management to change;
- Listen and act if a woman starts hinting there are problems, don't be dismissive because it makes you uncomfortable;
- Think broadly when asked to make suggestions of names for any position or role.

'useless' men fear it is a zero-sum game (in many ways it is) and seeing more women succeed means fewer men like them progress. Our society, overall, seems to be heading towards more polarization rather than less.

Nevertheless, the push for equity across the board, far beyond that of women in STEM, is so much in the public discourse it seems improbable to me that women will be pushed, literally or metaphorically, back into the kitchen. Whether all the women who *could* make fantastic breakthroughs or more modest contributions to our scientific and innovation world are supported in order to achieve their dreams remains to be seen. I would like to believe equity will win out over special interest groups wishing to maintain the *status quo*. Collectively, the issues are now well-known, the solutions are often obvious. The time has come when we must stop merely talking about the issues. For the benefit of all, we must build on our understanding and momentum to drive towards an inclusive society, breaking down the barriers that have held women back from entering science in the first place, or prospering when they do, for far too long.

ENDNOTES

Prelims

1. Quoted by Nancy Malkiel in *Keep the Damned Women Out* (Princeton University Press, Princeton, 2016).
2. Sometimes expanded to STEMM, with the second M corresponding to medicine. I will use STEM throughout, as my focus is on science.

Chapter 1

1. Rafaela von Bredow and Kerstin Kullmann, 'Women are just as gifted in science as men', *Spiegel International*, 20 August 2015, https://www.spiegel.de/international/germany/spiegel-interview-with-two-female-nobel-prize-recipients-a-1047838.html.
2. Michela Carlana, 'Implicit stereotypes: Evidence from teachers' gender bias', *Quarterly Journal of Economics*, 134, (2019): pp. 1163–224, doi: 10.1093/qje/qjz008.
3. Redrow Homes, *Poor Careers Advice Exacerbating the Construction Sector Skills-gap and Discouraging Apprentices*, 6 March 2017, https://www.redrow.co.uk/newsroom/national/2017/3/poor-careers-advice-and-misperceptions-are-exacerbating-the-construction-sector.
4. I will use quotes from other such Laureates throughout the book, leading scientists all, deriving them from a book 10 *Years of L'Oreal-Unesco* awards I was sent in 2009 when I myself won the European Laureate, celebrating the first ten years of their scheme. It does not seem to be widely available, although mini-biographies of all the winners can be found on the L'Oreal website, https://www.fondationloreal.com/our-programs-women-science/laureates-loreal-unesco-women-science-international-award.
5. Jo Bostock, *The Meaning of Success: Insights from Women at Cambridge*, (Cambridge University Press, Cambridge, 2014) and free online at https://www.cam.ac.uk/system/files/the_meaning_of_success_final_revised_for_print_final.pdf.

6. 'Give Me Inspiration! The Paradigm Shift', a series of conversations at Churchill College, https://www.chu.cam.ac.uk/about/events/conversations/.
7. Quoted by Nancy Hopkins in her introduction to Ben Barres' autobiography, *The Autobiography of a Transgender Scientist* (MIT Press, Boston, 2014).
8. Joshua Howgego, 'Jess Wade's one-woman mission to diversify Wikipedia's science stories', *New Scientist*, 5 February 2020, https://www.newscientist.com/article/mg24532680-800-jess-wades-one-woman-mission-to-diversify-wikipedias-science-stories/.
9. Mona Becker and Melanie Nilsson, 'College chemistry textbooks fail on gender representation', *Journal of Chemical Education*, 98, pp. 1146–51, (2021), doi: 10.1021/acs.jchemed.0c01037.
10. Wellcome Report, 'Young people's views on science education: Science Education Tracker 2019 Wave 2', https://cms.wellcome.org/sites/default/files/science-education-tracker-2019.pdf.
11. Karen Thomas, 'Donna Strickland: Girls just wanna have fun—in the lab', *S.P.I.E.*, 7 February 2019, https://spie.org/news/donna-strickland_girls-just-wanna-have-fun-%E2%80%93-in-the-lab?SSO=1.

Chapter 2

1. Mary Somerville, *Personal Recollections from Early Life to Old Age, of Mary Somerville, with Selections from Her Correspondence*, published by her daughter Martha Somerville: (John Murray, London, 1873).
2. Francis Crick, *What Mad Pursuit*, (Basic Books, New York, 1988), p. 69.
3. Chris Green, 'Could you name more than one female scientist?', *Independent*, 17 May 2014, https://www.independent.co.uk/news/science/could-you-name-more-than-one-female-scientist-9391307.html.
4. Mary Beard, *Women and Power: A Manifesto*, (Profile Books, London, 2017).
5. Chen Jie, 'Women and science do mix—and mix well', *China Daily*, 3 July 2008, http://www.chinadaily.com.cn/cndy/2008-03/07/content_6515745.htm.
6. Samuel Pepys, *The Diary of Samuel Pepys*, Vol 8, 1667, eds Robert Latham and William Matthews (Harper Collins, 1974).

7. Richard Holmes, *This Long Pursuit*, (William Collins, London, 2016).
8. Margaret Cavendish, *Observations upon Experimental Philosophy. To which is added, The Description of a New Blazing World*, (A. Maxwell, London, 1666).
9. Emma Wilkins, 'Margaret Cavendish and the Royal Society', *Notes and Records of the Royal Society*, 68, (2014): pp. 245–60, doi:10.1098/rsnr.2014.0015.
10. Samuel Mintz, 'The Duchess of Newcastle's visit to the Royal Society', *J. Engl. Germanic Philol.*, 51, (1952): pp. 168–76, https://www.jstor.org/stable/27713402.
11. Emma Wilkins, ibid.
12. Those who we call scientists today would have been known as natural philosophers at that time.
13. Patricia Fara, *Pandora's Breeches*, (Pimlico, London, 2004), p. 93.
14. Caroline Herschel *Memoir and Correspondence of Caroline Herschel*, ed. Mrs John Herschel, (London, John Murray, 1876): http://digital.library.upenn.edu/women/herschel/memoir/memoir.html#II.
15. Ibid.
16. Ibid.
17. Michael Hoskin, 'Herschel's 40ft reflector: funding and functions', *Journal for the History of Astronomy*, 34, (2003): pp. 1–32, doi:10.1177/002182860303400101.
18. Caroline Herschel, *Memoir and Correspondence of Caroline Herschel*, ed. Mrs John Herschel, (London, John Murray, 1876): http://digital.library.upenn.edu/women/herschel/memoir/memoir.html#II.
19. Quoted in Jan Golinski, *Science as Public Culture: Chemistry and Enlightenment in Britain, 1760–1820* (Cambridge University Press, Cambridge, 1992), p. 242.
20. Bence Jones, *The Royal Institution*, Project Gutenberg digital version https://www.gutenberg.org/files/46869/46869-h/46869-h.htm, p. 138.
21. John Herschel, *Review of Mechanism of the Heavens, Quarterly Review*, 47 (1832): pp. 537–59.
22. 'Sir, a woman's preaching is like a dog's walking on his hind legs. It is not done well; but you are surprised to find it done at all.'
23. Mary Somerville, *Personal Recollections from Early Life to Old Age, of Mary Somerville, with Selections from her Correspondence*, published by her daughter Martha Somerville: (London, John Murray, 1873).
24. James Edward Austen-Leigh, *A Memoir of Jane Austen*, (Richard Bentley, London, 1869).

25. Mary Somerville, *Personal Recollections from Early Life to Old Age, of Mary Somerville, with Selections from her Correspondence*, published by her daughter Martha Somerville: (London, John Murray, 1873).
26. Anonymous review, 'Mary Somerville', *Nature*, (1874): pp. 417–18.
27. Mary Somerville *Personal Recollections from Early Life to Old Age, of Mary Somerville, with Selections from her Correspondence*, published by her daughter Martha Somerville: (London, John Murray, 1873).
28. Tom Almeroth-Williams, 'Ahead of her time', University of Cambridge News story, 8 March 2021, https://www.cam.ac.uk/stories/mary-astell-collection-magdalene-college.
29. Catherine Sutherland, 'Books owned by Mary Astell in the Old Library of Magdalene College Cambridge', to appear in *The Library: The Transactions of the Bibliographical Society* (Forthcoming).
30. P.S. Laurie, 'The buildings and old instruments of the Royal Observatory, Greenwich', *The Observatory*, 80 (1960): pp. 13–22, https://articles.adsabs.harvard.edu/full/1960Obs....80...13L.
31. Ruth Perry, 'Mary Astell and the Enlightenment', Chapter 6.1 in *Women, Gender and the Enlightenment*, eds. Sarah Knott and Barbara Taylor (Palgrave MacMillan, London, 2005).
32. It would be more correct to say that she came up with the idea that such a program could be written, with the rudiments of an algorithm laid out.
33. Letter from Ada King to Mary Somerville, Somerville Papers, Oxford, Bodleian Libraries, Dep c. 367, Folder MSBY-3, fols 55v-56t and quoted in Christopher Hollings, Ursula Martin and Adrian Rice, *Ada Lovelace: The Making of a Computer Scientist*, (Oxford, Bodleian Library, 2018), p. 32.
34. Benjamin Woolley, *Ada Lovelace, Bride of Science*, (London, Pan Books, 2000), p. 259.
35. *Sketch of the Analytical Engine invented by Charles Babbage. L.F. Menabrea. With notes upon the Memoir by the translator, Ada Augusta Lovelace*, (R. and J.E. Taylor, London, 1843).
36. Desmond King-Hele, *Erasmus Darwin: A life of unparalleled achievement*, (London, Giles de la Mare Publishers, 1999), pp. 282–3.
37. Natasha Gelling, 'The women who mapped the Universe and still couldn't get any respect', *Smithsonian Magazine*, 18 September 2013, https://www.smithsonianmag.com/history/the-women-who-mapped-the-universe-and-still-couldnt-get-any-respect-9287444/.
38. Nancy Malkiel, *Keep the Damned Women out*, (Princeton University Press, Princeton, 2016).

39. I discussed this instance of Hertha Ayrton's candidature being rejected with Sarah-Jayne Blakemore in a 2015 video prepared for the Royal Society, exploring the issue of women entering the Society: https://www.youtube.com/watch?v=YIFO9dg83MA .
40. Institution of Engineering and Technology Archives Biographies: 'Hertha Ayrton' https://www.theiet.org/membership/library-archives/the-iet-archives/biographies/hertha-ayrton/.
41. Joan Mason, 'Hertha Ayrton (1854–1923) and the admission of women to the Royal Society of London', *Notes and Records of the Royal Society of London*, 45, (1991): pp. 201–20, doi: 10.1098/rsnr.1991.0019.
42. Hodgkin was an Oxford X-ray crystallographer and 1964 Nobel Prize winner for her work unravelling the structure of various biomolecules, including the vitamin B12 and, later, insulin.
43. Caroline Swash, 'Women Stained Glass Artists', *Arts and Crafts Tours Blog*, 12 June 2020, https://artsandcraftstours.com/blog/women-stained-glass-artists.
44. Quoted in Carroll Pursell, 'Am I a lady or an engineer? The origins of the Women's Engineering Society in Britain 1918-40', *Technology and Culture*, 34 (1993): pp. 78–97, https://www.jstor.org/stable/pdf/3106456.pdf.
45. Quoted in 'This month in physics history: 23 March 1882: Birth of Emmy Noether', *APS News* 22(3), March 2013, https://www.aps.org/publications/apsnews/201303/physicshistory.cfm.
46. Churchill College, of which I am Master, hold the personal archives of Lise Meitner, and a brief account of her life can be found at https://www.chu.cam.ac.uk/news/archives-centre/lise-meitner-max-planck-medal/.
47. Quoted by Ruth Sime in *Lise Meitner: A Life in Physics* (Oakland, University of California Press, 1996).
48. Ibid.
49. This and much more can be found discussed in the recordings of a symposium held at Churchill College in January 2019 and with which I was involved, celebrating the 80th Anniversary of the publication of her work on nuclear fission, https://www.chu.cam.ac.uk/news/news-and-events/lise-meitner/.
50. Quoted by Sandeep Ravindran in 'Barbara McClintock and the discovery of jumping genes', *PNAS* 109 (50), (2012): pp. 20198–9, doi:10.1073/pnas.1219372109.
51. Royal Society, Minutes of Council, 1920–26, 12, p. 180, minute 16, July 1923.

52. Joan Mason, 'The admission of the first women to the Royal Society of London', *Notes Rec. R. Soc. Lond.* 46, (1992): pp. 279–300, doi:10.1098/rsnr.1992.0027.
53. Eve Curie, *Madam Curie – a Biography*, (Doubleday, 1937).
54. Linda Lear, *Beatrix Potter: The Extraordinary Life of a Victorian Genius*, (Penguin, London, 2007).
55. Rosalind Franklin Institute is situated on the Harwell Campus in Oxfordshire, England, https://www.rfi.ac.uk/.
56. Royal Society Rosalind Franklin Award and Lecture: https://royalsociety.org/grants-schemes-awards/awards/rosalind-franklin-award/.
57. The Hahn Meitner Institute of the Helmholtz Zentrum, Berlin, https://www.helmholtz-berlin.de/zentrum/forschungszentrum/campus/historie/lise-meitner-campus/index_en.html.
58. The Lise-Meitner-Haus (Department of Physics) at Humboldt-Universität, Zu Berlin, https://www.physik.hu-berlin.de/en/department/about/the-lise-meitner-haus-department-of-physics/das-institutsgebaeude.
59. James Watson, *The Double Helix*, (W&N Reprint Edition, 2010).
60. Matilda Joslyn Gage, 'Woman as an inventor', *The North American Review*, 136, (1883): pp. 478–89, https://www.jstor.org/stable/25118273.
61. Susan Dominus, 'Women scientists were written out of history. It's Margaret Rossiter's lifelong mission to fix that', *Smithsonian Magazine*, October 2019, https://www.smithsonianmag.com/science-nature/unheralded-women-scientists-finally-getting-their-due-180973082/.
62. Westminster Gazette, 14 March 1909 quoted by Joan Mason in *Notes and Records of the Royal Society of London*, 45, (1991): pp. 201–20, https://www.jstor.org/stable/531699.
63. Parity relates to the symmetry of an elementary particle's wave function, the expression which describes the state of the particle. Parity can be even or odd. It was, prior to these experiments, expected to be conserved during the interactions between particles.
64. 'Give me inspiration! The paradigm shift with Dame Jocelyn Bell Burnell', https://www.youtube.com/watch?v=C44XKTHEwEo.
65. American Institute of Physics Oral History of Physics interview with Jocelyn Bell Burnell: https://www.aip.org/history-programs/niels-bohr-library/oral-histories/31792.
66. Max Perutz, Address at the memorial service for Dorothy Hodgkin in the University Church, Oxford, 4 May 1995, quoted in Georgina Ferry's biography of Hodgkin, *Dorothy Hodgkin: A Life*, (Granta Books, 1998, London), p. 171.

67. Ibid., pp. 289–90.
68. Margot Lee Shetterly, *Hidden Figures*, (Williams Collins, 2016).
69. Erin Cech, 'The intersectional privilege of white able-bodied heterosexual men in STEM', *Sci. Adv.*, 8 (24), (1922): eabo155, doi: 10.1126/sciadv.abo1558.
70. Amelia Horgan, *Mind the Gap: Student Experiences of Sexism within Teaching and Learning at Cambridge University*, (2015), https://drive.google.com/file/d/0BojbQADDLP9ATFFXMHQyZ3Z3RoU/view?resourcekey=0-4f-5ZdO8TOW9txnoCYeS3Q.
71. Jason McBride, 'Nobel laureate Donna Strickland: "I see myself as a scientist, not a woman in science"', *The Guardian*, 20 October 2018, https://www.theguardian.com/science/2018/oct/20/nobel-laureate-donna-strickland-i-see-myself-as-a-scientist-not-a-woman-in-science.
72. Vikas Shah, 'On gender & science – a conversation with May-Britt Moser, Nobel Prize Winning Scientist', *ThoughtEconomics*, 27 July 2017, https://thoughteconomics.com/may-britt-moser-interview/.
73. Felicia Taylor, 'Google's Marissa Mayer: Passion is a gender neutralizing force', *CNN Business*, 5 April 2012, https://edition.cnn.com/2012/04/05/tech/google-marissa-mayer/index.html.
74. Peter Medawar, *Advice to a Young Scientist*, (Basic Books, NY,1979), p. 23.
75. Bas Hofstra, Vivek Kulkarni, Sebastian Munoz-Najar Galvez, Bryan He, Dan Jurafsky, and Daniel McFarland, 'The diversity-innovation paradox', *PNAS*, 117, (2020): pp. 9284–91, doi:10.1073/pnas.1915378117.
76. Erin Cech, 'The intersectional privilege of white able-bodied heterosexual men in STEM', *Sci. Adv.*, 8 (24), eabo155, 1–14, doi: 10.1126/sciadv.abo1558.
77. Gendered Innovations in Science, Health and Medicine, Engineering and Environment, https://genderedinnovations.stanford.edu/.
78. Suk Kyeong Lee, 'Sex as an important biological variable in biomedical research', *BMB Reports*, 51 (2018): pp. 167–73, doi: 10.5483/BMBRep.2018.51.4.034.
79. Caroline Criado Perez, *Invisible Women*, (Penguin Random House, London, 2019).
80. 'Give me inspiration – the paradigm shift with Professor Dame Carol Robinson', 2015, https://soundcloud.com/churchill-college/give-me-inspiration-the-paradigm-shift-with-professor-dame-carol-robinson.
81. Angela Wright, Elisabeth Michielsens, Sylvia Snijders, Leena Kumarappan, Michele Williamson, Linda Clarke, and Peter Urwin, *Diversity in STEMM: Establishing a business case, Westminster Business School for the*

Royal Society's Diversity Programme 2014, https://royalsociety.org/-/media/policy/projects/leading-the-way/diversity-business-case-june-2014.pdf.

82. Ibid.

Chapter 3

1. Claudia Dreifus, 'Fighting sexism in science: Five questions for Meg Urry', *Undark*, 19 July 2018, https://undark.org/2018/07/19/five-questions-meg-urry/.
2. Peter Medawar, *The Art of the Soluble*. First published 1967; reissued 2021 (Routledge, Abingdon, 2021), p. 132.
3. Anthea Lipsett, 'Reality check', *The Guardian*, 2 June 2009, https://www.theguardian.com/education/2009/jun/02/interview-wendy-hall.
4. Vivian Hunt, Lareina Yee, Sara Prince, and Sundiatu Dixon-Fyle, 'Delivering through diversity', 18 January 2018, https://www.mckinsey.com/business-functions/organization/our-insights/delivering-through-diversity.
5. Quoted by Rebecca Casterton, *Train of Brain blog*, 'The incredible shyness of Henry Cavendish', 11 April 2014, https://trainofbrain.wordpress.com/2014/04/11/1-the-incredible-shyness-of-henry-cavendish/.
6. Sarah Gilbert and Catherine Green, *Vaxxers*, (Hodder and Stoughton, London, 2021), p. 6.
7. Quoted in Hannah Ellis-Petersen, 'Katie Bouman: the 29-year-old whose work led to first black hole photo', *The Guardian*, 11 April 2019, https://www.theguardian.com/science/2019/apr/11/katie-bouman-black-hole-photo.
8. Paul Nurse, 'Scientific collaboration', *Nobel Prize Inspiration Initiative*, May 2011, https://www.youtube.com/watch?v=Zzl6HKA3rrc&list=PLc8JX9rbkVougQZLwMqKloZ8CUzxvwIPe&index=5.
9. Françoise Barré-Sinoussi, 'What is your outlook on collaboration?', *Nobel Prize Inspiration Initiative*, 15 October 2020, https://www.youtube.com/watch?v=LEyj14Uz35s&list=PLc8JX9rbkVougQZLwMqKloZ8CUzxvwIPe&index=9&t=0s.
10. Henrique Pinheiro, Matt Durning, and David Campbell, 'Do women undertake interdisciplinary research more than men, and do self-citations bias observed differences?', *Quantitative Science Studies*, 3(2), pp. 363–92 (2022), doi: 10.1162/qss_a_00191.

11. Veronica Strang and Tom McLeish, 'Evaluating Interdisciplinary Research: a practical guide', Durham University Institute of Advanced Study, July 2015, https://www.iasdurham.org/wp-content/uploads/2020/11/StrangandMcLeish.EvaluatingInterdisciplinaryResearch.July2015_2.pdf.
12. Charles Darwin, letter to Joseph Hooker, 1 August 1857. 'Letter no. 2130'. *Darwin Correspondence Project*, https://www.darwinproject.ac.uk/letter/?docId=letters/DCP-LETT-2130.xml&query=2130.
13. Science Council, '10 types of scientists', https://sciencecouncil.org/about-science/10-types-of-scientist/.
14. Karthik Ramaswamy, 'Interview: Venki Ramakrishnan and the race to decode the mother of all molecules', 27 January 2019, *The Wire*, https://thewire.in/the-sciences/interview-venki-ramakrishnan-and-the-race-to-decode-the-mother-of-all-molecules.
15. Ephraim Hardcastle, 'Newsnight's Guardian-trained editor, Ian Katz, is keen on diversity', 19 March 2014, *Daily Mail*, https://www.dailymail.co.uk/debate/article-2583930/EPHRAIM-HARDCASTLE-Dan-Snow-blasts-Prince-Charles-memos-ministers.html.
16. Vanessa Friedman, 'Lessons of a Rosetta scientist', *New York Times*, 19 November 2014, https://www.nytimes.com/2014/11/20/fashion/the-lessons-of-a-rosetta-scientists-shirt.html.
17. Rafaela von Bredow and Kerstin Kullmann, 'Women are just as gifted as men', *Spiegel International*, 20 August 2015, https://www.spiegel.de/international/germany/spiegel-interview-with-two-female-nobel-prize-recipients-a-1047838.html.
18. Asia Game Changer Awards: Jean Liu, 1 November 2017, https://asiasociety.org/asia-game-changers/jean-liu.
19. International Finance Corporation, World Bank Group, Case Study Didi Chuxing inclusive workplace programs, 2020, https://www.ifc.org/wps/wcm/connect/4754b28a-7ab4-4d64-b3cf-fa78c2e70dc0/202008_D2E_DidiChuxing.pdf?MOD=AJPERES&CVID=nfvwCQW.
20. Margaret Thatcher, Speech to UN General Assembly (Global Environment), 8 November 1989, https://www.margaretthatcher.org/document/107817.
21. Royal Society Policy Briefing, 'The research and technical workforce in the UK', February 2021, https://royalsociety.org/-/media/policy/Publications/2021/2021-02-12-research-and-technical-workforce-in-the-uk.pdf.

22. Claudia Dreifus interviews Martin Rees 'On the best use of science to safeguard humanity', *Quanta Magazine*, 5 December 2018, https://www.quantamagazine.org/martin-rees-on-the-future-of-science-and-humanity-20181205//.
23. Jim Smith, 'The cell cycle and beyond: an interview with Paul Nurse', *Dis.Model Mech.* 2, (2009): pp. 113–15, doi: 10.1242/dmm.002592.
24. Rebeccao Trager, 'In situ with Frances Arnold', *Chemistry in Britain*, 1 April 2018, https://www.chemistryworld.com/culture/in-situ-with-frances-arnold/3008732.article.

Chapter 4

1. Alan Benson, 'Executive suite: advice from the top interviews, Sally Ride', *USA Today*, 19 March 2006, https://usatoday30.usatoday.com/money/companies/management/2006-03-19-sally-ride_x.htm.
2. Jocelyn Bell Burnell, American Institute of Physics Oral History Interviews 21 May 2000, https://www.aip.org/history-programs/niels-bohr-library/oral-histories/31792.
3. Frances Henry, *Survey of Women in the Academies of America*, May 2015, https://www.ancefn.org.ar/user/files/SURVEY_OF_WOMEN.pdf.
4. Rebecca Ratcliffe and Claire Shaw, '"Philosophy is for posh, white boys with trust funds"—why are there so few women?', *The Guardian*, 5 January 2015, https://www.theguardian.com/higher-education-network/2015/jan/05/philosophy-is-for-posh-white-boys-with-trust-funds-why-are-there-so-few-women.
5. Lin Bian, Sarah-Jane Leslie, and Andrei Cimpian, 'Gender stereotypes about intellectual ability emerge early and influence children's interests', *Science*, 355, (2017): pp. 389–91, doi: 10.1126/science.aah6524.
6. David Miller, Kyle Nolla, Alice Eagly, and David Uttal, 'The development of children's gender-science stereotypes: A meta-analysis of 5 decades of U.S. draw-a-scientist studies', *Child Development*, 89, (2018), pp. 1943–55, doi:10.1111/cdev.13039.
7. Nick Chambers, Jordan Rehill, Elnaz Kashefpakdel, and Christian Percy, 'Drawing the future' for Education and Employers, January 2018, https://www.educationandemployers.org/wp-content/uploads/2018/01/DrawingTheFuture.pdf.
8. Louise Archer, Jennifer DeWitt, Jonathan Osborne, Justin Dillon, Beatrice Willis, and Billy Wong, '"Doing" science versus "being" a scientist: Examining 10/11-year-old schoolchildren's constructions of science

through the lens of identity', *Science Education*, 94, (2010): pp. 617–39, doi: 10.1002/sce.20399.

9. Institute of Mechanical Engineers Report, 'When STEM? A question of age', https://www.imeche.org/policy-and-press/reports/detail/when-stem-a-question-of-age.
10. Adam Withall, 'Lego letter from the 1970s still offers a powerful message to parents 40 years later', *Independent*, 23 November 2014, https://www.independent.co.uk/news/lego-letter-from-the-1970s-still-offers-a-powerful-message-to-parents-40-years-later-9878303.html.
11. Geena Davis Institute on Gender in Media Lego Creativity Study, September 2021, https://seejane.org/wp-content/uploads/LEGO-Ready-for-Girls-Creativity-Study.pdf.
12. Helen Russell, 'Lego to remove gender bias from its toys after findings of child survey', *The Guardian*, 11 October 2021, https://www.theguardian.com/lifeandstyle/2021/oct/11/lego-to-remove-gender-bias-after-survey-shows-impact-on-children-stereotypes.
13. Laura David, 'Training effects on mental rotation, spatial orientation and spatial visualisation depending on the initial level of spatial abilities', *Procedia—Social and Behavioral Sciences*, 33, (2012): pp. 328–32, doi: 10.1016/j.sbspro.2012.01.137.
14. Jonathan Wai, David Lubinski, and Camilla Benbow, 'Spatial ability for STEM domains: Aligning over 50 years of cumulative psychological knowledge solidifies its importance', *Journal of Educational Psychology*, 101, (2009): pp. 817–35, doi: org/10.1037/a0016127.
15. Let Toys be Toys Report, *Who Gets to Play? What Do Toy ads on UK TV Tell Children about Boys' and Girls' Play?*, (2015), http://lettoysbetoys.org.uk/wp-content/uploads/2015/12/LetToysBeToys-Advertising-Report-Dec15.pdf.
16. Mary Beth Quirk, 'Computer engineer Barbie needs men to write code, can't reboot computer', *Consumerist*, 18 November 2014, https://consumerist.com/2014/11/18/computer-engineer-barbie-needs-men-to-write-code-cant-reboot-computer/.
17. Rachel Weber, 'Mattel apologises for sexist developer Barbie book', *Gamesindustry.biz*, 21 November 2014, https://www.gamesindustry.biz/articles/2014-11-21-mattel.
18. Nicola Slawson, 'Vaccinologist Barbie: Prof Sarah Gilbert honoured with a doll', *The Guardian*, 4 August 2021, https://www.theguardian.com/society/2021/aug/04/vaccinologist-barbie-prof-sarah-gilbert-honoured-with-a-doll and *Barbie Role Models: Sarah Gilbert*, Mattel website, https://shopping.mattel.com/en-gb/pages/barbie-role-models-sarah-gilbert.

19. Sarah Gilbert and Catherine Green, *Vaxxers*, (Hodder and Stoughton, London, 2021), p. 230.
20. Institute of Physics Report, *It's Different for Girls*, October 2012, https://www.iop.org/sites/default/files/2019-04/its-different-for-girls.pdf.
21. Institute of Physics Report, *Why Not Physics?*, May 2018, https://www.iop.org/sites/default/files/2018-10/why-not-physics.pdf.
22. Marina Bassi, Mercedes Mateo Díaz, Rae Lesser Blumberg, and Ana Reynoso, 'Failing to notice? Uneven teachers' attention to boys and girls in the classroom', *IZA Journal of Labor Economics*, 7: 9 (2018), doi: 10.1186/s40172-018-0069-4.
23. Emma Burns, Keiko Bostwick, Rebecca Collie, and Andrew Martin, 'Understanding girls' disengagement: identifying patterns and the role of teacher and peer support using latent growth modeling', *J. Youth and Adolescence*, 48: pp. 979–95 (2019), doi: 10.1007/s10964-019-00986-4.
24. Institute of Physics, 'Inclusive Teaching: 10 tips for teachers', https://spark.iop.org/collections/inclusive-teaching.
25. 'Give me inspiration! The paradigm shift with Melanie Welham', https://www.youtube.com/watch?v=gBuI7IS7BSU.
26. Henry Nicholls, '"You must have the courage of your convictions": A conversation with Jane Goodall', *Pacific Standard*, 17 April 2014, https://psmag.com/environment/must-courage-convictions-conversation-jane-goodall–79229.
27. Institute of Physics Report, *Closing Doors*, December 2013, https://www.iop.org/sites/default/files/2019-03/closing-doors.pdf.
28. Sylvie Kerger, Romain Martin, and Martin Brunner, 'How can we enhance girls' interest in scientific topics?', *British Journal of Educational Psychology*, 81, (2011): pp. 606–62, doi:10.1111/j.2044-8279.2011.02019.x.
29. Becky Francis, Louise Archer, Julie Moore, Jen de Will, and Lucy Yeomans, 'Femininity, science, and the denigration of the girly girl', *British J of Sociology of Education*, 38, (2017): pp. 1097–110, doi: 10.1080/01425692.2016.1253455.
30. Diana Betz and Denise Sekaquaptewa, 'My fair physicist? Feminine math and science role models demotivate young girls', *Social Psychological and Personality Science*, 3, (2012): pp. 738–46, doi: 10.1177/1948550612440735.
31. Kevin Rask and Elizabeth Bailey, 'Are faculty role models? Evidence from major choice in an undergraduate institution', *J. Econ. Education*, 33, (2002): pp. 99–124, doi: 10.1080/00220480209596461.

32. Jiesi Guo, Herbert Marsh, Philip Parker, Theresa Dicke, and Brooke van Zanden, 'Countries, parental occupation, and girls' interest in science', *The Lancet*, 393, (2019): E6-8, doi: 10.1016/S0140-6736(19)30210-7.
33. Gijsbert Stoet and David Geary, 'The gender-equality paradox in science, technology, engineering, and mathematics education', *Psychological Science*, 29, (2018): pp. 581–93, doi: 10.1177/0956797617741719; Corrigendum (2019) doi: 10.1177/0956797619892892.
34. Johanna Mellén and Petra Angervall, 'Gender and choice: differentiating options in Swedish upper secondary STEM programmes', *J. Education Policy*, 36, (2020): pp. 417–35, doi: 10.1080/02680939.2019.1709130.
35. Claude Steele, *Whistling Vivaldi* (W.W. Norton and Company, New York, 2010).
36. Jenessa Shapiro and Amy Williams, 'The role of stereotype threats in undermining girls'and women's performance and interest in STEM fields', *Sex Roles*, 66, (2012): pp. 175–83, doi: 10.1007/s11199-011-0051-0.
37. Shen Zhang, Toni Schmader and William Hall, 'L'eggo my ego: reducing the gender gap in math by unlinking the self from performance', *Self and Identity*, 12 (2013): pp. 400–12, doi: 10.1080/15298868.2012.687012.
38. Esra Çetinkaya, Sarah D. Herrmann, and Yasemin Kisbu-Sakarya, 'Adapting the values affirmation intervention to a multi-stereotype threat framework for female students in STEM', *Social Psychology of Education*, 23, (2020): pp. 1587–607, doi:10.1007/s11218-020-09594-8.
39. Amanda Diekman, Emily Clark, and Aimee Belanger, 'Finding common ground: Synthesizing divergent theoretical views to promote women's STEM pursuits', *Social Issues and Policy Review*, 113, (2019): pp. 182–210, doi: 0.1111/sipr.12052.
40. Carol Dweck, *Mindset: The New Psychology of Success*, (New York, Random House 2006).
41. Carol Dweck, 'What having a "growth mindset" actually means', *Harvard Business Review*, 13 January 2016, https://hbr.org/2016/01/what-having-a-growth-mindset-actually-means.

Chapter 5

1. Mae Jemison, *Teach Arts and Sciences Together*, TED2002, February 2002, https://www.ted.com/talks/mae_jemison_teach_arts_and_sciences_together?language=en.

2. Wendy Jewell, *Fabiola Gianotti—a "My Hero" Story, Secrets of the Universe*, 14 January 2014, http://secretsoftheuniversefilm.com/__trashed/.
3. Paul Nurse, 'There's a relationship between creativity and humour', *Nobel Prize Inspiration Initiative*, 8 June 2016, https://www.youtube.com/watch?v=sVUDKYBtgY4&list=PLc8JX9rbkVoug14ofvx8ebrZyUTx6uK8i&index=3.
4. Simon Baron-Cohen, *The Essential Difference*, (London, Penguin, 2012).
5. David Greenberg, Varun Warrier, Carrie Allison, and Simon Baron-Cohen, 'Testing the Empathizing–Systemizing theory of sex differences and the Extreme Male Brain theory of autism in half a million people', *PNAS*, 115, (2018): pp. 12152–7, doi: 10.1073/pnas.1811032115.
6. Simon Baron-Cohen, *The Pattern Seekers* (London, Penguin, 2022).
7. Gina Rippon, *The Gendered Brain*, (London, Bodley Head, 2019).
8. Cordelia Fine, *Delusions of Gender*, (London, Icon Books, 2010).
9. Cordelia Fine, *Testosterone Rex*, (London, Icon Books, 2017).
10. Angela Saini, *Inferior*, (London, 4th Estate, 2017).
11. Cordelia Fine, 'Is there neurosexism in functional neuroimaging investigations of sex differences?', *Neuroethics*, 6, (2013): pp. 369–409, doi: 10.1007/s12152-012-9169-1.
12. Daphna Joel and Ricardo Tarrasch, 'On the mis-presentation and misinterpretation of gender-related data: The case of Ingalhalikar's human connectome study', *PNAS*, 111, E637 (2014). Doi: 10.1073/pnas.1323319111.
13. Anelis Kaiser, Sven Haller, Sigrid Schmitz, and Cordula Nitsch, 'On sex/gender related similarities and differences in fMRI language research', *Brain Research Reviews*, 61(2), (2009): pp. 49–59, doi: 10.1016/j.brainresrev.2009.03.005.
14. Lisa Eliot, Adnan Ahmed, Hiba Khan, and Julie Patel, 'Dump the "dimorphism": comprehensive synthesis of human brain studies reveals few male-female differences beyond size', *Neuroscience and Biobehavioral Reviews*, 125: pp. 667–97 (2021), doi: 10.1016/j.neubiorev.2021.02.026.
15. Giorgio Innocenti and David Price, 'Exuberance in the development of cortical networks', *Nature Reviews Neuroscience*, 6, (2005): pp. 955–65, doi: 10.1038/nrn1790.
16. Fatima Ismail, Ali Fatemi, and Michael Johnston, 'Cerebral plasticity: Windows of opportunity in the developing brain', *European Journal of Paediatric Neurology*, 21(1), (2017): pp. 23–48, doi: 10.1016/j.ejpn.2016.07.007.
17. Charles Zeanah, Charles Nelson, Nathan Fox, Anna Smyke, Peter Marshall, Susan Parker, and Sebastian Koga, 'Designing research

to study the effects of institutionalization on brain and behavioral development: The Bucharest Early Intervention Project', *Development and Psychopathology*, 15, (2003): pp. 885–907, doi: 10.1017/s0954579403000452.

18. Bryan Kolb and Robbin Gibb, 'Brain plasticity and behaviour in the developing brain', *J. Can. Acad. Child Adolescent Psychiatry*, 20, (2011): pp. 265–76, doi:10.1016/B978-0-444-63327-9.00005-9.
19. Eleanor Maguire, David Gadian, Ingrid Johnsrude, and Christopher Frith, 'Navigation-related structural change in the hippocampi of taxi drivers', *PNAS*, 97, (2000): pp. 4398–403, doi: 10.1073/pnas.070039597.
20. Katherine Woollett, Hugo Spiers, and Eleanor Maguire, 'Talent in the taxi: a model system for exploring expertise', *Philosophical Transactions of the Royal Society B: Biological Sciences*, 364(1522), (2009): pp. 1407–16, doi: 10.1098/rstb.2008.0288.
21. Eva-Maria Griesbauer, Ed Manley, Jan Wiener, and Hugo Spiers, 'London taxi drivers: A review of neurocognitive studies and an exploration of how they build their cognitive map of London', *Hippocampus*, 32(1), (2022): pp. 3–20, doi: 10.1002/hipo.23395.
22. Alice Lee, 'Data for the problem of evolution in man. VI.—A first study of the correlation of the human skull', *Proc. Roy. Soc.*, 671, (1901): pp. 435–41, doi: 10.1098/rspl.1900.0038.
23. Gina Rippon, Lise Eliot, Sarah Genon, and Daphna Joel, 'How hype and hyperbole distort the neuroscience of sex differences', *PLoS Biology*, 19(5), (2021): e3001253, doi: 10.1371/journal.pbio.3001253.
24. Gina Rippon, Rebecca Jordan-Young, Anelis Kaiser, and Cordelia Fine, 'Recommendations for sex/gender neuroimaging research: key principles and implications for research design, analysis, and interpretation', *Front. Hum. Neurosci.*, 8, Art. 650 (2014), doi: 10.3389/fnhum.2014.00650.
25. Lise Eliot, Adnan Ahmed, Hiba Khan, and Julie Patel, 'Dump the "dimorphism": Comprehensive synthesis of human brain studies reveals few male-female differences beyond size', *Neuroscience & Biobehavioral Reviews*, 125, (2021): pp. 667–97, doi: 10.1016/j.neubiorev.2021.02.026.
26. Jim Smith, 'The cell cycle and beyond: an interview with Paul Nurse', *Disease Models Mechanisms*, 2, (2009): pp. 113–15, doi: 10.1242/dmm.002592.
27. James Byrnes, David Miller, and W Schafer, 'Gender differences in risk taking: A meta-analysis', *Psychological Bulletin*, 125, (1999): pp. 367–83, doi:10.1037/0033-2909.125.3.367.

28. Eike Weber, Ann-Renee Blais, and Nancy Betz, 'A domain-specific risk-attitude scale: measuring risk perceptions and risk behaviors', *J. Behavioral Decision Making*, 15, (2002): pp. 263–90, doi: 10.1002/bdm.414.
29. Margo Gardner and Laurence Steinberg, 'Peer influence on risk taking, risk preference, and risky decision making in adolescence and adulthood: An experimental study', *Developmental Psychology*, 41, (2005): pp. 625–35, doi: 10.1037/0012-1649.41.4.625.
30. Sarah-Jayne Blakemore, *Inventing Ourselves*, (London, Penguin Random House, 2018).
31. James Byrnes, David Miller, and William Schafer, 'Gender differences in risk taking: A meta-analysis', *Psychological Bulletin*, 125, (1999): pp. 367–73, doi: 10.1037/0033-2909.125.3.367.
32. Venki Ramakrishnan, *Gene Machine*, (London, Oneworld Publications Ltd, 2018), p. 91.
33. Fabiola Gianotti, 'Doctor Fabiola Gianotti on being the first female director of the greatest fundamental science experiment on earth', *Humans of Science*, 1 February 2018, https://www.humans-of-science.org/single-post/2018/01/30/Doctor-Fabiola-Gianotti-on-being-the-first-female-director-of-the-greatest-fundamental-science-experiment-on-earth.
34. Allison Master and Andrew Meltzoff, 'Cultural stereotypes and sense of belonging contribute to gender gaps in STEM', *International Journal of Gender, Science and Technology*, 12(1), (2020): pp. 152–98, https://genderandset.open.ac.uk/index.php/genderandset/article/view/674.
35. A toast allegedly raised by Keats to Newton at the 'immortal dinner party' hosted by Benjamin Haydon in 1817, as reported, for instance, in Denise Gigante, *Publications of the Modern Language Association*, 117, (2002): pp. 433–48, stable URL: https://www.jstor.org/stable/823143.
36. David Parrish, Creative Industries definitions, https://www.davidparrish.com/creative-industries-definitions/.
37. Peter Medawar, 'Science and Literature' Romanes Lecture, published in *Perspectives in Biology and Medicine*, 12, (1969): pp. 529–46, doi: 10.1353/pbm.1969.0005.
38. Cristina Odone, 'My daughter shouldn't have to study science', *Daily Telegraph*, 7 November 2015, https://www.telegraph.co.uk/education/educationopinion/11978621/My-daughter-shouldnt-have-tostudy-science-says-Cristina-Odone.html.
39. Hansard, Lords Debate: Careers Education for Students Volume 792: debated on Thursday 6 September 2018, https://hansard.parliament.uk/lords/2018-09-06/debates/93FB7039-9CBE-4058-AABA-659546E318E0/CareersEducationForStudents.

40. Christopher Frayling, 'Research in art and design', *Royal College of Art Research Papers*, 1, (1993/4): pp. 1–5, https://researchonline.rca.ac.uk/384/3/frayling_research_in_art_and_design_1993.pdf.
41. Tim Brown, *Change by Design*, (New York, Harper Business, 2009).
42. As quoted by Matthew Reisz, 'The core connection', *Times Higher Education*, 7 January 2010, https://www.timeshighereducation.com/features/the-core-connection/409838.article.
43. Claire Maniez, 'An interview with Siri Hustvedt', *Transatlantica*, 2, (2016), doi: 10.4000/transatlantica.8328.
44. Richard Rhodes, *Hedy's Folly*, (New York, Doubleday, 2011).
45. Magdolna Hargittal, *Women Scientists*, (Oxford, Oxford University Press, 2015), p. 132.
46. Royal Society, *Vision for Science and Mathematics Education*, 2014, https://royalsociety.org/-/media/education/policy/vision/reports/vision-full-report-20140625.pdf.
47. C.P. Snow, *The Two Cultures*, 50th Anniversary Printing (Cambridge, Cambridge University Press, 1998) p. 15.
48. 'Give me inspiration! The paradigm shift with Professor Nancy Rothwell', 30 October 2019, https://www.youtube.com/watch?v=2EgC2TWyZEs.

Chapter 6

1. Mathias Döpfner, 'Interview: BioNTech founders Özlem Türeci and Ugur Sahin on developing the BioNTech-Pfizer COVID-19 vaccine, the future of fighting cancer, and whether people can live to 200', *Business Insider*, 23 March 2021, https://www.businessinsider.com/pfizer-biontech-vaccine-creators-science-covid-19-dose-health-2021-3?r=US&IR=T.
2. Rita King, 'Weizmann scientist and Nobel Prize Laureate Ada Yonath, has four tips for young people interested in a career in science', *The Curiosity Review*, 27 August 2014, https://www.weizmann-usa.org/blog/weizmann-scientist-and-nobel-laureate-ada-yonath-has-four-tips-for-young-people-interested-in-a-career-in-science/ .
3. Fermi Award Winners: Q&A; Dr Mildred S. Dresselhaus, *US Department of Energy, 6 June* 2012, https://web.archive.org/web/20150908034322/https://science.energy.gov/news/featured-articles/2012/06-06-12.
4. See, for instance, the book Fabrizio Butera and John M. Levine eds., *Coping with Minority Status: responses to inclusion and exclusion* edited by Fabrizio

Butera and John Levine (New York: Cambridge University Press, 2009), doi: 10.1017/CBO9780511804465.

5. Self-efficacy is a term introduced by the US-Canadian psychologist Albert Bandura in 1977 (Albert Bandura, 'Self-efficacy: towards a unifying theory of behavioural change', *Psychological Review*, 84, (1977): pp. 191–215, doi: 10.1037/0033-295X.84.2.191). It is defined as an individual's belief in their capacity to execute the necessary behaviours to produce specific performance attainments.
6. Xiaoxia Huang, Jie Zhang, and Laura Hudson, 'Impact of math self-efficacy, math anxiety, and growth mindset on math and science career interest for middle school students: the gender moderating effect', *Eur. J. Psychology of Education*, 34 (2019): pp. 621–40, doi: 10.1007/s10212-0180-0403-z.
7. Una Telhed, Martin Bäckström, and Fredrik Björklund, 'Will I fit in and do well? The importance of social belongingness and self-efficacy for explaining gender differences in interest in STEM and HEED majors', *Sex Roles*, 77 (2017): pp. 86–96, doi: 10.1007/s11199-0016-0694-y.
8. UNESCO data for Sustainable Development, Institute for Statistics Blog 'Data + Policy = Action on International Day for Women and Girls in Science', 14 May 2021, http://uis.unesco.org/en/blog/data-policy-action-international-day-women-and-girls-science.
9. Ruth van Veelen, Belle Derks, and Maaike Dorine Endedijk, 'Double trouble: How being outnumbered and negatively stereotyped threatens career outcomes of women in STEM', *Frontiers in Psychology*, 10:150 (2019), doi: 10.3389/fpsyg.2019.00150.
10. Kathi Miner, Samantha January, Kelly Dray, and Adrienne Carter-Sowell, 'Is it always this cold? Chilly interpersonal climates as a barrier to the well-being of early-career women faculty in STEM', *Equality, Diversity and Inclusion*, 38 (2019): pp. 226–45, doi: 10.1108/EDI-07-2018-0127.
11. Valerie Bostwick and Bruce Weinberg, 'Nevertheless she persisted? Gender peer effects in doctoral STEM programs', *Journal of Labor Economics*, 40, (2022): pp. 397–436, doi: 10.1086/714921.
12. 'Give me inspiration: the paradigm shift with Professor Nancy Rothwell', 30 October 2019, https://www.youtube.com/watch?v=2EgC2TWyZEs.
13. Lihadh Al-Gazali in *10 Years of the L'Oreal-UNESCO Awards*, For Women in Science, L'Oreal Fondation Paris, (2008), pp. 106–7.
14. To assess field-specific ability beliefs, participants in the study were asked to rate their agreement with four statements concerning what is

required for success in their field, such as 'Being a top scholar of [discipline] requires a special aptitude that just can't be taught'. Respondents rated both the extent to which they personally agreed with these statements, and the extent to which they believed other people in their field would agree with the statements. Answers were averaged to produce the field-specific ability belief score for each subject, a higher score implying a greater emphasis on raw ability.

15. African Americans, like women, are stereotyped as lacking innate intellectual talent; not so Asian Americans.
16. Alysa Malespina, Christian Schunn, and Chandralekha Singh, 'Whose ability and growth matter? Gender, mindset and performance in physics', *Int. J. of STEM Education*, 9:28 (2022), doi: 10.1186/s40594-011-00342-2.
17. Emma Ideal and Rhiannon Meharchand, *Blazing the Trail; Essays by Leading Women in Science*, (2013), ISBN 1482709430.
18. TEDx Whitehall, Athene Donald, 'Nothing is Wasted: turning negative experiences into positive life lessons', January 2018, https://www.ted.com/talks/athene_donald_nothing_is_wasted_turning_negative_experiences_into_positive_life_lessons.
19. American Institute of Physics Centre for the History of Physics 2000–22, Marie Curie and the Science of Radioactivity exhibit, quote taken from *Marie Curie, Pierre Curie with Autobiographical Notes*. Translated by Charlotte and Vernon Kellogg (New York: Macmillan, 1923). https://history.aip.org/exhibits/curie/brief/06_quotes/quotes_03.html.
20. Robin McKie, 'Venkatraman Ramakrishnan interview: "It takes courage to tackle very hard problems in science," *The Guardian*, 24 November 2013, https://www.theguardian.com/science/2013/nov/24/venkatraman-ramakrishnan-ribosome-nobel-chemistry.
21. 'Give me inspiration! The paradigm shift with Dame Jocelyn Bell Burnell', 4 December 2018, https://www.youtube.com/watch?v=C44XKTHEwEo.
22. Matin Durrani, 'Donna Strickland gives inside story of her Nobel-prize-winning research', *Physics World*, 20 February 2019, https://physicsworld.com/a/donna-strickland-gives-inside-story-of-her-nobel-prize-winning-research/.
23. London School of Hygiene and Tropical Medicine, Women Leaders in Global Health, Professor Sally Davies—Women Leaders Insight, https://www.lshtm.ac.uk/research/research-action/women-leaders-global-health/insights-women-leaders/prof-dame-sally-davies.

24. Jo Bostock ed., *The Meaning of Success*, (Cambridge University Press, Cambridge, 2014), pp. 20–1.
25. Adam Gristwood and Holger Breithaupt, 'Science in politics: An interview with Dame Anne Glover, former Chief Scientific Advisor to the President of the European Commission', *EMBO Reports*, 20:e48349 (2019), doi: 10.15252/embr.201948349.
26. Pat O'Connor, Clare O'Hagan, Eva Sophia Myers, Liv Baisner, Georgi Apostolov, Irina Topuzova, Gulsun Saglamer, Mine G. Tan, and Hulya Caglayan, 'Mentoring and sponsorship in higher education institutions: men's invisible advantage in STEM?', *Higher Education Research & Development*, 39, (2020): pp. 764–77, doi: 10.1080/07294360.2019.1686468.
27. Adam Gristwood and Holger Breithaupt, 'Science in politics: an interview with Dame Anne Glover, former Chief Scientific Advisor to the President of the European Commission', *EMBO Reports*, 20:e48349 (2019), doi: 10.15252/embr.201948349.
28. Venki Ramakrishnan, *Gene Machine: The Race to Decipher the Secrets of the Ribosome*, (Harper Collins, 2018) p. 263.
29. Quoted in *Rosalind Franklin: The Dark Lady of DNA*, Brenda Maddox, (London, Harper Collins, 2002), p. 138.
30. Rafaela von Bredow and Kerstin Kullman, 'Women are just as gifted in science as men', 20 August 2015, *Der Spiegel*, https://www.spiegel.de/international/germany/spiegel-interview-with-two-female-nobel-prize-recipients-a-1047838.html.
31. Africa Villanueva-Felez, Richard Woolley, and Carolina Cañibano, 'Nanotechnology researchers' collaboration relationships: a gender analysis of access to scientific information', *Social Studies of Science*, 45 (2015): pp. 100–29, doi 10.1177/0306312714552347.
32. Xiao Han Zeng, Jordi Duch, Marta Sales-Pardo, João Moreira, Fillipi Radicchi, Haroldo Ribeiro, Teresa Woodruff, and Luis Amaral, 'Differences in collaboration patterns across discipline, career stage, and gender', *PLOS Biology*, 14(11): e1002573 (2016), doi:10.1371/journal.pbio.1002573.
33. Marek Kwiek and Wojciech Roszka, 'Gender-based homophily in research: A large-scale study of man-woman collaboration', *J. of Informetrics*, 15, (2021): 101171, doi: 10.1016/j.joi.2021.101171.
34. Bruno Latour and Steve Woolgar, *Laboratory Life: The Construction of Social Facts*, (Princeton University Press, Chichester, 1986).
35. Troy Heffernan, 'Sexism, racism, prejudice and bias: a literature review and synthesis of research surrounding student evaluations of courses

and teaching', *Assessment & Evaluation in Higher Education*, 47, (2022): pp. 144–54, doi: 10.1080/02602938.2021.1888075.

36. Lillian McNeil, Adam Driscoll, and Andrea Hunt, 'What's in a name: Exposing gender bias in student ratings of teaching', *Innovative Higher Ed.*, 40, (2015): pp. 291–303, doi 10.1007/s10755-014-9313-4.
37. Rate my Professors is a US site encouraging college students to comment on their teachers https://www.ratemyprofessors.com/.
38. Website: Gendered language in teacher reviews: http://benschmidt.org/profGender/#.
39. Margret Sigudardottir, Gudbjorg Rafnsdottir, Anna Jónsdóttir, and Dadi Kristofersson, 'Student evaluation of teaching: gender bias in a country at the forefront of gender equality', *Higher Education Research and Development*, (2022): pp. 1–14, doi: 10.1080/07294360.2022.2087604.
40. Merryn McKinnon and Christine O'Connell, 'Perceptions of stereotypes applied to women who publicly communicate their STEM work', *Humanities and Social Sciences Comms.*, 7:160 (2020), doi: /10.1057/s41599-020-00654-0.
41. Inoka Amarasekara and Will Grant, 'Exploring the YouTube science communication gender gap: A sentiment analysis', *Public Understanding of Science*, 28, (2018): pp. 68–84, doi: 10.1177/0963662518786654.
42. Ruth van Veelen, Belle Derks, and Maaike Dorine Endedijk, 'Double trouble: How being outnumbered and negatively stereotyped threatens career outcomes of women in STEM', *Frontiers in Psychology*, 10:150 (2019), doi: 10.3389/fpsyg.2019.00150.
43. S. Cheryan and H.R. Markus Masculine, 'Defaults: identifying and mitigating hidden cultural biases', *Psychological Review*, 127, (2020): pp. 1022–52, doi: 10.1037/rev0000209.

Chapter 7

1. Interview with Jocelyn Bell Burnell, American Institute of Physics Oral Histories: https://www.aip.org/history-programs/niels-bohr-library/oral-histories/31792.
2. Meg Urry, 'Diminished by discrimination we scarcely see', *Washington Post*, 6 February 2005, http://www.washingtonpost.com/wp-dyn/articles/A360-2005Feb5.html.
3. Venki Ramakrishnan, *Gene Machine*, (OneWorld Publications, London, 2019), p. 102.

4. NGCP: 'The state of girls and women in STEM', March 2022, https://ngcproject.org/sites/default/files/downloadables/2022-03/ngcp_stateofgirlsandwomeninstem_2022b.pdf.
5. WISE Campaign: Workforce statistics for 2019–20, https://www.wisecampaign.org.uk/updated-workforce-statistics-to-september-2020/.
6. Garcia Working Papers 5, *Academic Careers and Gender Inequality*, edited by Farah Dubois-Shaik and Bernard Fusulier, https://eige.europa.eu/sites/default/files/garcia_working_paper_5_academic_careers_gender_inequality.pdf.
7. Eileen Pollack, *The Only Woman in the Room*, (Beacon Press, Boston, 2015).
8. Corinne Moss-Racusin, John Dovidio, Victoria Brescol, Mark Graham, and Jo Handelsman, 'Science faculty's subtle gender biases favor male students', *PNAS*, 109 (41), (2012): pp. 16474–9, doi: 10.1073/pnas.1211286109.
9. The Royal Society, 15 June 2011, *Mothers in Science*, https://royalsociety.org/_/media/royal_society_content/about-us/equality/2011-06-15-mothers-in-science.pdf. This booklet was subsequently updated and expanded: *Parent, Carer, Scientist*, 7 March 2016, https://royalsociety.org/topics-policy/diversity-in-science/parent-carer-scientist/.
10. Eva Amsen, 'An interview with Ottoline Leyser', *Development*, 138(22), (November 2011): pp. 4815–17, doi: 10.1242/dev.075333.
11. Erin Cech and Mary Blair-Loy, 'The changing career trajectories of new parents in STEM', *PNAS*, 116(10) (2019): pp. 4182–7, doi: 10.1073/pnas.1810862116.
12. This means extending how long an assistant professor is allowed to prove themselves before being assessed for a permanent, i.e. tenured position.
13. Heather Antecol, Kelly Bedard, and Jenna Stearns, 'Equal but inequitable: who benefits from gender-neutral tenure clock stopping policies', *American Economic Review*, 108(9), (2018): pp. 2420–41, doi: 10.1257/aer.20160613.
14. Meg Urry, 'Diminished by discrimination we scarcely see', *Washington Post*, 6 February 2005, https://www.washingtonpost.com/archive/opinions/2005/02/06/diminished-discrimination-we-scarcely-see/9c69e9e6-c013-4f7c-a214-d410b0dbc565/.
15. John Biggers, 'Dame Anne McLaren obituary', *The Guardian*, 10 July 2007, https://www.theguardian.com/science/2007/jul/10/uk.obituaries.

16. Matthias Krapf, Heinrich Ursprung and Christian Zimmerman, 'Parenthood and productivity of highly skilled labor: Evidence from the groves of academe', *J. Econ. Behavior and Organization*, 140, (2017): pp. 147–75.
17. Gemma Derrick, Adam Jaeger, Pei-Ying Chen, Cassidy Sugimoto, Thed Van Leeuven, and Vincent Lariviere, 'Models of parenting and its effect on academic productivity: preliminary results from an international survey', *17th International conference on Scientometrics and Informetrics Vol II*, (2019), pp. 167–76, ISBN: 978-88-3381-118-5.
18. Eve Higginbotham and Maria Lund Dahlberg, Editors; *Impact of COVID-19 on the Careers of Women in Academic Sciences, Engineering, and Medicine*, National Academies Press (2021), ISBN 978-0-309-26837-0: doi 10.17226/26061.
19. Jordan Dworkin, Kristin Linn, Erin Teich, Perry Zurn, Russell Shinohara, and Danielle Bassett, 'The extent and drivers of gender imbalance in neuroscience reference lists.' *Nat. Neurosci.*, 23, (2020): pp. 918–26, doi: 10.1038/s41593-020-0658-y.
20. EPSRC Report, 'Understanding our Portfolio, A gender perspective', 2022, https://epsrc.ukri.org/files/aboutus/epsrcunderstandingourportfolio-agenderperspectivereport.
21. Royal Society of Chemistry Report, *Is publishing in the chemical sciences gender biased?*, (2019), https://www.rsc.org/new-perspectives/talent/gender-bias-in-publishing/.
22. Nyssa Silbiger and Amber Stubler, 'Unprofessional peer reviews disproportionately harm underrepresented groups in STEM.' *Peer J.*, 7:e8247 (2019), doi: 10.7717/peerj.8247.
23. Rita Colwell, *A Lab of One's Own*, (Simon and Schuster, NY, 2020), p. 1.
24. These are terms originally introduced in a different context by psychologists Keith Stanovich and Richard West, and discussed at length in Daniel Kahneman's book *Thinking Fast and Slow*, (paperback Penguin, London, 2012).
25. Doyin Atewologun, Tinu Cornish, and Fatima Tresh, *Unconscious bias training: an assessment of the evidence for effectiveness*, Equality and Human Rights Commission, Research Report 1132018, https://www.equalityhumanrights.com/sites/default/files/research-report-113-unconcious-bais-training-an-assessment-of-the-evidence-for-effectiveness-pdf.pdf.
26. Virginia Valian, *Why So Slow?*, (MIT Press, Cambridge MA, 1999).

27. L.S. Fidell, 'Empirical verification of sex discrimination in hiring practices in psychology', in *Woman: Dependent or independent variable*, pp. 774–82, eds. R.K. Unger and F.L. Denmark, (Psychological Dimensions, New York, 1975).
28. Corinne Moss-Racusin, John Dovidio, Victoria Brescol, Mark Graham, and Jo Handelsman, 'Science faculty's subtle gender biases favor male students', *PNAS*, 109 (41), (2012): pp. 16474–9, doi: 10.1073/pnas.1211286109.
29. Christine Wennerås and Agnes Wold, 'Nepotism and sexism in peer-review', *Nature*, 387, (1997): pp. 341–3, doi: 10.1038/387341a0.
30. Juliet Eilperin, 'White House women want to be in the room where it happens', *Washington Post*, 13 September 2016, https://www.washingtonpost.com/news/powerpost/wp/2016/09/13/white-house-women-are-now-in-the-room-where-it-happens/.
31. Video clip, Channel 4 News, 8 October 2020, https://www.facebook.com/Channel4News/videos/mr-vice-president-im-speaking/2733521616895439/.
32. 'An interview with Nancy Hopkins', *MIT's Infinite MIT project*, https://infinitehistory.mit.edu/video/nancy-h-hopkins.
33. Members of the First and Second Committees on Women Faculty in the School of Science, *A Study on the Status of Women Faculty in Science at MIT*, 1999, http://web.mit.edu/fnl/women/women.html#The%20Study.
34. *A Report on the Status of Women Faculty in the Schools of Science and Engineering at MIT*, 2011, https://news.mit.edu/sites/default/files/documents/women-report-2011.pdf.
35. Corinne Moss-Racusin, John Dovidio, Victoria Brescol, Mark Graham, and Jo Handelsman, 'Science faculty's subtle gender biases favor male students', *PNAS*, 109 (41) (2012): pp. 16474–9, doi: 10.1073/pnas.1211286109.
36. Asia Eaton, Jessica Saunders, Ryan Jacobson, and Keon West, 'How gender and race stereotypes impact the advancement of scholars in STEM: Professors' biased evaluations of physics and biology post-doctoral candidates', *Sex Roles*, 82, (2020): pp. 127–41, doi: 10.1007/s11199-019-01052-w.
37. Toni Schmader, Jessica Whitehead, and Vicki Wysocki 'A linguistic comparison of letters of recommendation for male and female chemistry and biochemistry job applicants', *Sex Roles*, 57(7–8) (2007): pp. 509–14, doi: 10.1007/s11199-007-9291-4.

38. Whitney Morgan, Katherine Elder, and Eden King, 'The emergence and reduction of bias in letters of recommendation', *Journal of Applied Social Psychology*, 43(11) (2013): pp. 2297–306, doi: 10.1111/jasp.12179.
39. Kuheli Dutt, Danielle Pfaff, Ariel Bernstein, Joseph Dillard, and Caryn Block, 'Gender differences in recommendation letters for postdoctoral fellowships in geoscience', *Nature Geoscience*, 9, (2016): pp. 805–8, doi: 10.1038/ngeo2819.
40. Tom Forth, Gender Bias Calculator, https://www.tomforth.co.uk/genderbias/.
41. Molly King, Carl Bergstrom, Shelley Correll, Jennifer Jacquet, and Jevin D. West, 'Men set their own cites high: gender and self-citation across fields and over time', *Socius: Sociological Research for a Dynamic World*, 3 (2017): pp. 1–22, doi: 10.1177/2378023117738903.
42. Vincent Larivière, Chaoqun Ni, Yves Gingras, Blaise Cronin, and Cassidy Sugimoto. 'Bibliometrics: global gender disparities in science', *Nature*, 504, (2013): pp. 211–13, doi: 10.1038/504211a.
43. Jordan Dworkin, Kristin Linn, Erin Teich, Perry Zurn, Russell Shinohara, and Danielle Bassett, 'The extent and drivers of gender imbalance in neuroscience reference lists', *Nat. Neurosci.*, 23 (2020): pp. 918–26, doi: 10.1038/s41593-020-0658-y.
44. The order of M (man) and W (woman) reflect the order in the author list; where there were more than two authors, their gender was not taken into account.
45. EPSRC Report, 'Understanding our Portfolio, A gender perspective', 2022, https://epsrc.ukri.org/files/aboutus/epsrcunderstandingourportfolio-agenderperspectivereport/.
46. Isabelle Vernos, 'Quotas are questionable', *Nature*, 495 (2013): p. 39, doi: 10.1038/495039a.
47. DFG Press Release No. 29, 1 July 2020, 'Equal opportunity in research: still significant need for action', https://www.dfg.de/en/service/press/press_releases/2020/press_release_no_29/index.html.
48. *The Metric Tide: Report of the Independent Review of the Role of Metrics in Research Assessment and Management*, July 2015, https://webarchive.nationalarchives.gov.uk/ukgwa/20210802101914/https://re.ukri.org/sector-guidance/publications/metric-tide/.
49. Meg Urry, 'Diminished by discrimination we scarcely see', *Washington Post*, 6 February 2005, https://www.washingtonpost.com/archive/opinions/2005/02/06/diminished-discrimination-we-scarcely-see/9c69e9e6-c013-4f7c-a214-d410b0dbc565/.

50. 'Sexual Harassment of Women: Climate, Culture and Consequences in Academic Sciences, Engineering and Medicine', National Academies of Sciences, Engineering and Medicine, http://nap.naptionalacademies.org/24994.
51. Kathryn Clancy, 'Transcript of my oral testimony from February 27th Congressional hearing on sexual misconduct in the sciences', 28 February 2018, https://kateclancy.com/transcript-of-my-oral-testimony-from-february-27th-congressional-hearing-on-sexual-misconduct-in-the-sciences/.
52. 'Updated statement: CERN stands for diversity', https://press.cern/news/press-release/cern/updated-statement-cern-stands-diversity.
53. Athene Donald, 'Every female scientist could produce a list of when she was treated differently or harassed', Wired, 5 October 2018, https://www.wired.co.uk/article/nobel-prize-2018-winners-women-science.
54. Public Affairs, UC Berkeley, 'A message about Professor Marcy's resignation', *Berkeley News*, 14 October 2015, https://news.berkeley.edu/2015/10/14/a-message-about-professor-marcys-resignation/.
55. Jocelyn Kaiser, 'Astronomer Geoff Marcy booted from National Academy of Sciences in wake of sexual harassment', *Science*, 27 May 2021, https://www.science.org/content/article/astronomer-geoff-marcy-booted-national-academy-sciences-wake-sexual-harassment.
56. 'Give me inspiration: The paradigm shift with Dame Sally Davies', interview between Sally Davies and Athene Donald, 4 July 2017, Churchill College, https://www.youtube.com/watch?v=54M2AlFS9Eo.
57. Kathryn Clancy, Robin Nelson, Julienne Rutherford, and Katie Hinde, 'Survey of Academic Field Experiences (SAFE): Trainees report harassment and assault', *PLoS One*, 9(7) (2014): e102172, doi: 10.1371/journal.pone.0102172.
58. Ibid.
59. AIP Oral Histories, David deVorkin interview with Jocelyn Bell Burnell, 21 May 2000, https://www.aip.org/history-programs/niels-bohr-library/oral-histories/31792.
60. Adrianne Appel, 'A passion for science without barriers: Nancy Hopkins, renowned champion of gender equality, looks back over her career', *Nature*, 481, (2012): p. 13, doi: 10.1038/481013a.

Chapter 8

1. Adam Gristwood and Holger Breithaupt, 'An interview with Dame Anne Glover, former Chief Scientific Advisor to the President of the European Commission', *EMBO Reports* 20: e48349 (2019), doi: 10.15252/embr.201948349.
2. Zoe Corbyn, 'Frances Arnold: "To expect a Nobel prize is rather silly"', *The Guardian*, 21 October 2018, https://www.theguardian.com/science/2018/oct/21/frances-arnold-interview-ability-engineer-most-complicated-things-on-planet-nobel-prize-enzymes-chem.
3. *Gender Equality in Science: Inclusion and Participation of Women in Global Science Organizations. Results of two global surveys.* September 2021, Interacademies Partnership, https://www.interacademies.org/sites/default/files/2021-10/Gender%20Equality%20in%20Science.pdf ISBN: 978889440446.
4. 50:50 The equality project, https://www.bbc.co.uk/5050/partners/home/.
5. The Life Scientific podcasts, https://www.bbc.co.uk/programmes/b015sqc7/episodes/downloads.
6. Mona Becker and Melanie Nilsson, 'College chemistry textbooks fail on gender representation', *J. Chem. Educ.*, 98, 4, (2021): pp. 1146–51, doi: 10.1021/acs.jchemed.0c01037.
7. Geena Davis Institute on Gender in Media, 'The Scully effect: I want to believe in STEM', 2018, https://seejane.org/research-informs-empowers/the-scully-effect-i-want-to-believe-in-stem/.
8. Geena Davis Institute on Gender in Media, 'Representations of women STEM characters in media', 2021, https://seejane.org/research-informs-empowers/portray-her/.
9. Margot Lee Shetterly, *Hidden Figures* (William Collins, 2017).
10. Peter Nash and Emily Lomax, 'Managing extended paternity leave', Chartered Institute of Personnel and Development, December 2020, https://www.cipd.co.uk/Images/managing-extended-paternity-leave_tcm18-88297.pdf.
11. House of Commons Science and Technology Committee, Oral evidence: Diversity and inclusion in STEM, HC 903, 27 April 2022,

https://committees.parliament.uk/event/13461/formal-meeting-oral-evidence-session/.

12. House of Commons Science and Technology Committee, Oral evidence: Diversity and inclusion in STEM, HC 903, 18 May 2022, https://committees.parliament.uk/oralevidence/10228/html/. I had also made an earlier written submission, https://committees.parliament.uk/writtenevidence/42294/pdf/, and followed up with additional data, https://committees.parliament.uk/writtenevidence/109051/pdf/.
13. Devi Sridhar, *Preventable*, (Viking, Milton Keynes, 2022).
14. Once the Department took over the digital brief, after my time as chair, I assume the workforce changed, but at the time it was very noticeable.
15. Royal Society Report, 'Vision for science and mathematics education', 2014, https://royalsociety.org/-/media/education/policy/vision/reports/vision-full-report-20140625.pdf.
16. Martha Bottia, Elizabeth Stearns, Roslyn Mickleson, Stephanie Moller, and Lauren Valentino, 'Growing the roots of STEM majors: female math and science high school faculty and the participation of students in STEM', *Economics of Education Review*, 45 (2015): pp. 14–27, doi:1016/j.econedurev.2015.01.002.
17. Richard Larson, Navid Ghaffarzadegan, and Yi Xue, 'Too many PhD graduates or too few academic job openings: The basic reproductive number R_0 in academia', *Sys. Res. Behav.*, 31(6), (2014): pp. 745–50, doi: 10.1002/sres.2210.
18. By which I mean labs not simply devoted to product development but researching in far less constrained ways, searching for totally new ideas and concepts.
19. Frank Dobbin and Alexandra Kalev, 'Why doesn't diversity training work?', *Anthropology Now*, 10 (2018): pp. 48–55, doi: 10.1080/19428200.2018.149318.
20. Still a college which admits only women as students.
21. Jill Armstrong, *Women Collaborating with Men: Inclusive networking and sponsorship*, Murray Edwards College, University of Cambridge (2019), https://www.murrayedwards.cam.ac.uk/sites/default/files/files/Inclusive%20Networking%20and%20Sponsorship_FINAL.pdf.
22. Jill Armstrong and Jason Ghaboos, *Women Collaborating with Men: Everyday workplace inclusion*, Murray Edwards College, University of Cambridge (2019), https://www.murrayedwards.cam.ac.uk/sites/default/files/files/Everyday%20Workplace%20Inclusion_FINAL.pdf.

23. Sarah Teichmann, Muslifah Haniffa, and Jasmin Fisher, 'Community voices: policy proposals to promote inclusion in academia through the lens of women in science', *Nat. Commun.*, 13, (2022): p. 4068, doi: 10.1038/s41467-022-31616-6.
24. Danielle Gaucher, Justin Friesen, and Aaron C. Kay, 'Evidence that gendered wording in job advertisements exists and sustains gender inequality', *Journal of Personality and Social Psychology*, 101(1), (2011): pp. 109–28, doi: 10.1037/a0022530.
25. Shenggang Hu, Jabir Alshehabi Al-Ani, Karen Hughes, Nicole Denier, Alla Konnikov, Lei Ding, Jinhan Xie, Yang Hu, Monideepa Tarafdar, Bei Jiang, Linglong Kong, and Hongsheng Dai, 'Balancing gender bias in job advertisements with text-level bias mitigation', *Frontiers in Big Data*, 5, (2022): p. 805713, doi: 10.3389/fdata.2022.805713.
26. Athene Donald, 'UKRI job advert: conspiracy or cock-up?', *Times Higher Education*, 6 February 2017, https://www.timeshighereducation.com/blog/ukri-job-advert-conspiracy-or-cock.
27. 'Give me inspiration! The paradigm shift with Dame Jocelyn Bell Burnell', https://www.youtube.com/watch?v=C44XKTHEwEo.
28. National Academies of Sciences, Engineering, and Medicine, *The impact of Covid-19 on the careers of women in academic sciences, engineering, and medicine*, The National Academies Press (2021), doi: 10.17226/26061.
29. Mark Anderson, 'Mildred Dresselhaus: The Queen of Carbon', *IEEE Spectrum*, 28 April 2015, https://spectrum.ieee.org/geek-life/profiles/mildred-dresselhaus-the-queen-of-carbon.
30. Athene Donald, 'Just one action for women in science', *The Guardian*, 19 June 2015, https://www.theguardian.com/science/occams-corner/2015/jun/19/just-one-action-for-women-in-science.
31. Pilita Clark, 'Women must demand the right to be as useless as men', *Financial Times*, 24 February 2019, https://www.ft.com/content/bb565e54-35c5-11e9-bd3a-8b2a211d90d5.

FIGURE AND TABLE CREDITS

Figure 1 Mary Astell's annotations in her copy of René Descartes *Les Principes de la Philosophie*. Photo by permission of the Master and Fellows of Magdalene College, Cambridge.

Figure 2 Distribution of men and women between disciplines, aggregated across the Americas' academies. Frances Henry, *Survey of Women in the Academies of America*, May 2015, https://www.ancefn.org.ar/user/files/SURVEY_OF_WOMEN.pdf. Used with permission.

Figure 3 Global proportion of women and men in science as graduates and researchers, 2018. UNESCO data for Sustainable Development, Institute for Statistics Blog 'Data + Policy = Action on International Day for Women and Girls in Science', 14 May 2021, http://uis.unesco.org/en/blog/data-policy-action-international-day-women-and-girls-science. Used with permission.

Figure 4 Percentage of women at different career stages for US universities for biological and life sciences, student data from 2014 and academic data from 2015. 'Gender equity: Addressing recruitment at the departmental level', *Inside eLife*, 1 November 2018, https://elifesciences.org/inside-elife/6118bb63/gender-equity-addressing-recruitment-at-the-departmental-level. Data from (USA) National Center for Science and Engineering Statistics, 2017. Used with permission.

Figure 5 Field-specific ability beliefs and the percentage of female 2011 American PhDs in A) STEM and B) Social Science and Humanities. Sarah-Jane Leslie, Andrei Cimpian, Meredith Meyer, and Edward Freeland, 'Expectations of brilliance underlie gender distributions across academic disciplines', *Science* 347 (2015): pp. 262–7, doi: 10.1126/science.1261375. Used with permission.

Figure 6 Mean competence score given to male and female applicants by the MRC reviewers as a function of their scientific productivity, measured as total impact. Christine Wennerås and Agnes Wold, 'Nepotism and sexism in peer-review', *Nature* 387 (1997): pp. 341–3, doi: 10.1038/387341a0. Used with permission.

Figure 7 1988 *Punch* cartoon by Riana Duncan, 8 January 1988. Used with permission.

Figure 8 Degree of over- and under-citation of different author genders within MM and W∪W reference lists. Jordan Dworkin, Kristin Linn, Erin Teich, Perry Zurn, Russell Shinohara, and Danielle Bassett, 'The extent and drivers of gender imbalance in neuroscience reference lists.' *Nature Neuroscience* 23 (2020): pp. 918–26 doi: 10.1038/s41593-020-0658-y. Used with permission.

Figure 9 The public consciousness of sexual harassment and specific sexually harassing behaviours, much of which is 'below the water line'. National Academies of Sciences, Engineering, and Medicine, 2018. *Sexual Harassment of Women: Climate, Culture, and Consequences in Academic Sciences, Engineering, and Medicine*. Washington, DC: The National Academies Press. https://doi.org/10.17226/24994. Used with permission.

Figure 10 Data from UNESCO Fact Sheet 60, 'Women in Science'. Magdelena Szmigiera, 'Share of female research & development (R&D) researchers in 2017, by world region', *Statista*, 21 January 2022.

Table 1 Percentage of women members of those national academies which have at least 20% female members. *Gender Equality in Science: Inclusion and Participation of Women in Global Science Organizations. Results of Two Global Surveys.* September 2021, Interacademies Partnership. https://www.interacademies.org/sites/default/files/2021-10/Gender%20Equality%20in%20Science.pdf.

Table 2 Athene Donald, 'Just one action for women in science', *The Guardian*, 19 June 2015, https://www.theguardian.com/science/occams-corner/2015/jun/19/just-one-action-for-women-in-science.

PUBLISHER'S ACKNOWLEDGEMENTS

We are grateful for permission to include the following copyright material in this book:

Chapter 2

Extract from Francis Crick, *What Mad Pursuit*, (Basic Books, New York, 1988), p. 69.

Chapter 3

Extract from Claudia Dreifus, 'Fighting Sexism in science: Five questions for Meg Urry', *Undark*, 19 July 2018, https://undark.org/2018/07/19/five-questions-meg-urry/.

Extract from Peter Medawar, *The Art of the Soluble*, (Routledge, Abingdon, 1967, 2021), p. 132.

Extract from Anthea Lipsett, 'Reality check', *The Guardian*, 2 June 2009, https://www.theguardian.com/education/2009/jun/02/interview-wendy-hall.

Chapter 4

Extract from Alan Benson 'Executive suite: Advice from the top interviews, Sally Ride', *USA Today*, 19 March 2006, https://usatoday30.usatoday.com/money/companies/management/2006-03-19-sally-ride_x.htm.

'Quote from Jocelyn Bell Burnell, American Institute of Physics Oral History Interviews', 21 May 2000, https://www.aip.org/history-programs/niels-bohr-library/oral-histories/31792.

Chapter 5

Quote from Mae Jemison, 'Teach arts and sciences together', TED2002, February 2002, https://www.ted.com/talks/mae_jemison_teach_arts_and_sciences_together.

Quote from Fabiola Gianotti in *Secrets of the Universe* (2019), http://secretsoftheuniversefilm.com.

Quote from Paul Nurse, 'There's a relationship between creativity and humour', Nobel Prize Inspiration Initiative, 8 June 2016, https://www.youtube.com/watch?v=sVUDKYBtgY4&list=PLc8JX9rbkVoug14ofvx8ebrZyUTx6uK8i&index=3.

Chapter 6

Extract from 'Interview: BioNTech founders Özlem Türeci and Ugur Sahin on developing the BioNTech-Pfizer COVID-19 vaccine, the future of fighting cancer, and whether people can live to 200', *Business Insider*, 23 March 2021, https://www.businessinsider.com/pfizer-biontech-vaccine-creators-science-covid-19-dose-health-2021-3?r=US&IR=T.

Extract from Rita J. King, 'Weizmann scientist and Nobel Prize Laureate Ada Yonath, has four tips for young people interested in a career in science', *The Curiosity Review*, 27 August 2014, https://www.weizmann-usa.org/blog/weizmann-scientist-and-nobel-laureate-ada-yonath-has-four-tips-for-young-people-interested-in-a-career-in-science/.

Extract from 'Fermi Award Winners: Q&A; Dr Mildred S. Dresselhaus', US Department of Energy, 6 June 2012, https://web.archive.org/web/20150908034322/https://science.energy.gov/news/featured-articles/2012/06-06-12.

Chapter 7

Quote from Jocelyn Bell Burnell, American Institute of Physics Oral Histories: https://www.aip.org/history-programs/niels-bohr-library/oral-histories/31792.

Extract from Meg Urry, 'Diminished By Discrimination We Scarcely See', *Washington Post*, February 6 2005, http://www.washingtonpost.com/wp-dyn/articles/A360-2005Feb5.html.

Extract from Venki Ramakrishnan, *Gene Machine,* (OneWorld Publications, London, 2019), p. 102.

Chapter 8

Extract from Adam Gristwood and Holger Breithaupt, 'Science in politics: An interview with Dame Anne Glover, former Chief Scientific Advisor to the President of the European Commission', *EMBO Reports,* 20: e48349 (2019), https://www.embopress.org/doi/full/10.15252/embr.201948349.

Extract from Zoë Corbyn, 'Frances Arnold: "To expect a Nobel prize is rather silly"', *The Guardian,* 21 October 2018, https://www.theguardian.com/science/2018/oct/21/frances-arnold-interview-ability-engineer-most-complicated-things-on-planet-nobel-prize-enzymes-chem.

INDEX